Mushroom Technology

Second Edition

Lt. S Rajan MSc, PhD

Assistant Professor
Department of Microbiology
MR Government Arts College
Mannargudi

N Sivakumar MSc, PhD

Assistant Professor
Department of Microbiology
School of Biotechnology
Madurai Kamaraj University
Madurai

CBS

CBS Publishers and Distributors Pvt Ltd

New Delhi • Bengaluru • Chennai • Kochi • Kolkata • Mumbai
Bhopal • Bhubaneswar • Hyderabad • Jharkhand • Nagpur • Patna
• Pune • Uttarakhand • Dhaka (Bangladesh) • Kathmandu (Nepal)

Mushroom Technology

Second Edition

ISBN: 978-93-89565-77-5

Second Edition: 2020

First Edition: 2019

Published by Satish Kumar Jain and produced by Varun Jain for

CBS Publishers and Distributors Pvt Ltd

4819/XI Prahlad Street, 24 Ansari Road, Daryaganj, New Delhi 110 002, India.

Ph: 23289259, 23266861, 23266867   Website: www.cbspd.com

Fax: 011-23243014   e-mail: delhi@cbspd.com; cbspubs@airtelmail.in.

Corporate Office: 204 FIE, Industrial Area, Patparganj, Delhi 110 092

Ph: 011-4934 4934   Fax: 011-4934 4935   e-mail: publishing@cbspd.com;

publicity@cbspd.com

Branches

- **Bengaluru:** Seema House 2975, 17th Cross, K.R. Road, Banasankari 2nd Stage, Bengaluru 560 070, Karnataka
  Ph: +91-80-26771678/79   Fax: +91-80-26771680   e-mail: bangalore@cbspd.com
- **Chennai:** 7, Subbaraya Street, Shenoy Nagar, Chennai 600 030, Tamil Nadu
  Ph: +91-44-26680620, 26681266   Fax: +91-44-42032115   e-mail: chennai@cbspd.com
- **Kochi:** 68/1534, 35, 36 Power House Road, Opp. KSEB, Kochi 682018, Kerala
  Ph: +91-484-4059061-65   Fax: +91-484-4059065   e-mail: kochi@cbspd.com
- **Kolkata:** 6/B, Ground Floor, Rameswar Shaw Road, Kolkata-700 014, West Bengal
  Ph: +91-33-22891126, 22891127, 22891128   e-mail: kolkata@cbspd.com
- **Mumbai:** 83-C, Dr E Moses Road, Worli, Mumbai-400018, Maharashtra
  Ph: +91-22-24902340/41   Fax: +91-22-24902342   e-mail: mumbai@cbspd.com

Representatives

Bhopal	0-8319310552	Bhubaneswar	0-9911037372	Hyderabad	0-9885175004
Jharkhand	0-9811541605	Nagpur	0-9421945513	Patna	0-9334159340
Pune	0-9623451994	Uttarakhand	0-9716462459	Dhaka	01912-003485
Kathmandu	977-9818742655			(Bangladesh)	
(Nepal)					

Printed at India Binding House, Noida, UP, India

Preface

Mushrooms are the edible fleshy fruiting bodies of certain fungi, which may be gathered wild or grown under cultivation. The most commonly cultivated mushroom species are *Agaricus bisporus, Pleurotus* sp. and *Valvaria* sp. Although many other species are now gaining recognition in India due to the widespread consumption of Asian countries. Mushrooms contain a lot of proteins and minerals, several B vitamins and are regarded as a healthy food or food supplement. Moreover, due to certain chemical compounds valued for their medicinal properties, mushrooms gain more and more interest from the health food industry.

The book *Mushroom Technology* contains 35 chapters. This book describes: The history of mushroom cultivation in India, morphology and life cycle of mushroom, classification of mushroom, edible and poisonous mushrooms, isolation and maintenance of mushroom, cultivation of different types of mushrooms, nutritive value of mushrooms, post-harvest handling, mushroom recipes, mushroom diseases, abnormalities and competitor mushrooms, insect and pest of mushrooms, economics of mushroom cultivation and medicinal values of mushrooms.

We hope that this book will be very helpful to the mushroom farmers, researchers, students of different disciplines from various universities. We earnestly welcome constructive criticisms and valuable suggestions from the book users. Readers may send suggestions to ksrajan99@gmail.com

Lt. S Rajan
N Sivakumar

Contents

Chapter 1

INTRODUCTION

Mushroom is a large heterogeneous group of fungi. They showed different shapes, size and colours. They are quite different from each other in their characters, requirement of temperature and substrate, appearance and edibility. There are more than 2000 edible species of mushrooms in the world. Purkayastha and Chandra (1976) have listed 139 edible mushroom species available in India. Mushrooms are the fruiting bodies of macrofungi. They include both edible/medicinal and poisonous species. However, originally, the word "mushroom" was used for the edible members of macrofungi and "toadstools" for poisonous ones of the "gill" macrofungi. Scientifically the term "toadstool" has no meaning at all and it has been dropped altogether in order to avoid confusion and the terms edible, medicinal and poisonous mushrooms are used.

Edible mushrooms once called the "food of the gods" have now be taken regularly as part of the human diet or be treated as healthy or functional food. The extractable products from medicinal mushrooms, designed to supplement the human diet not as regular food, but as the enhancement of health and fitness. This can be classified into dietary supplements and mushroom nutriceuticals. Dietary supplements are ingredients extracted from foods, herbs, mushrooms and other plants that are taken without further modification for their presumed health-enhancing benefits.

Mushrooms are the most priced commodity among vegetables, not only because of its nutritive value but also for its characteristic aroma and flavour. They are very nutritious products that can be generated from lignocellulosic waste materials and are in rich in crude fibre and protein. Infact, mushrooms also contain low fat, low calories and good vitamins. In addition, many mushrooms possess multi-functional medicinal properties.

There is one Chinese statement that "Medicines and foods have a common origin". Mushrooms constitute a most rapidly growing new food category as SCP (Single cell protein). It contains 45% of proteins, which is higher than any other vegetable protein, but it is low in calories. It also contains higher amount of vitamins and minerals like calcium, phosphorous, iron, potassium and copper which is more essential for formation of bones and teeth. Another important feature is that mushrooms contains low amount of starch, which is good for peoples suffering from diabetes.

Mushrooms lack chlorophyll and hence they cannot get their energy from the sun through photosynthesis. Instead, during their vegetative growth stage, mushroom mycelia secrete enzymes that break down compounds such as cellulose and lignin present in the substrate. The degraded compounds are then absorbed by the hyphae and the mycelium enlarges, and in some cases growing several meters in diameter within the substrate.

Environmental factors such as temperature and light that are stimulating the second or reproductive growth stage. Cells of one mycelial strain fuse with cells of the opposite type to form a mycelium that contains both types of nuclei. The new mycelium continue to grow and eventually develop into a fruiting body. The gills of which are lined with spore bearing cells called basidia. Various mechanisms trigger the dispersal of spores, which in turn lodge in a substrate, become hyphae and begin the new cycle.

Chapter 2

HISTORY OF MUSHROOM PRODUCTION

Mushrooms have been considered as delicious food for several thousand years. It is one of the most important horticulture crops. Traditionally people used the wild collected mushrooms according to their personal knowledge of edible and poisonous varieties.

The first common mushroom cultivated was *Agaricus* in France during 1550-1650. The cultivation of white button mushroom was started first in France (1630) in the open ridge made out of horse dung. In 1650, Bonnexons a French man at Royal Academy of Sciences mentioned about the compost preparation and mushroom cultivation. Cultivation technology of this mushroom was passed from France to England and then to USA in the late 19th century.

Earliest description of how to grow mushroom was written by De Tournefort, Published in Paris in 1707. During latter period, French mushroom specialist and Mycologists innovated the improved cultural practices for mushroom cultivation, which were being followed till today.

- In 1731, Miller introduced French method of mushroom cultivation.
- In 1779, Ahercrombie described a method of composting stable manure in stacks.
- A problem of mushroom cultivation was rectified by two French Mycologists Matrocot and Constantina in 1894. They used sulphur fumigation and sterilization technique to reduce fungal and bacterial contamination.
- In 1902, Miss Ferguson illustrated the methodology of spore germination. '
- In 1905, Duggar from USA invented Cell suspension culture from mushroom tissue.
- During 1915 mushroom growers in USA introduced a second stage of composting called 'sweat in gout'.
- In 1918, USA marketed pure culture of bottled spawns. It was produced by spores from mushroom germinating them under aseptic conditions and then injecting into a bottle with sterilized compost.
- In 1918, Lambert (USA) produced pure culture of spawn in the glass bottles and began to sell to the mushroom growers.
- In India, Bose (1921) was successfully cultured two Agaricus mushroom on a sterilized dung media.
- In 1932, Sinden discovered the use of grain in making spawn and patented.
- In 1936, Pizer indicated the role of gypsum in compost, which stimulates the growth of mycelium rapidly.

- In 1941, Padwick reported successful cultivation of A. bisporus from various countries.
- In 1945, Mushroom growers association was started by mushroom grower of England and Ireland.
- In 1950, Edward produced formula for synthetic compost.
- In 1950, Sinden and Hauser prepared synthetic compost consisting of wheat straw, urea and other fertilizers.
- 1n 1962, Rasmusen (Denmark) produced startling results with pig manure.
- Cultivation of paddy straw mushroom might have started about 200 years ago in China but was recorded only in 1822 as being grown in a mandatory in Kwangtung, China. During 1935 this mushroom was introduced in the Philippines, Malaysia and other South East Asian countries.

Now mushrooms are being cultivated in morethan 100 countries with an estimated total production of over 5 million tones.

Chapter 3

HISTORY OF MUSHROOM CULTIVATION IN INDIA

India is one among the 100 countries who uses mushroom as a delicious food. Initial references for mushroom cultivation in India were probably in 1886.

- In India, N.W. Neuton (1886) was exhibited the mushroom growth at the annual show of Agri-Horticulture Security of India.
- In 1896-97, Dr. H.C. Roy of Calcutta Medical College analyzed the contents of mushrooms collected from caves and mines.
- In 1908, Sir David Prain initiated edible mushrooms cultivation by thorough search of edible mushrooms in India.
- In 1918, Kirtikar of Indian Medical Devices recorded some mushrooms.
- In 1921, Prof. S. R. Bose from Bengal succeeded in cultivining two species of *Agaricus* on a sterilized dung Media.
- In 1943, Thomas and his group from Coimbatore cultivated straw mushroom successfully.
- In 1962, Bano et al., obtained increased yield of *Pleurotus* sp. on paddy straw.
- In 1947, Asthara used flour of gram for the cultivation of *Volvariella*.

During the early 1950's the Government of Himachal Pradesh (HP) appointed Dr.S. Jain as the first Plant Pathologist and Mycologist. He worked in the Wild Flower Hall in Chharabra, Shimla. He was touring the interior areas of Himachal to help the apple orchardists and the farmers to control the diseases of apples, other fruits and crops like potatoes and wheat. While staying with some farmers in interior areas he noticed that there were rotting twigs on branches of apple and other fruit trees and wheat straw in the barn along with cow dung. In the suitable environmental conditions there were a profusion of mushrooms growing in the dark barns. This led him to think of using the waste material with the farmers for growing edible mushrooms. He made a research proposal on growing of edible mushrooms and got the permission for the same from the State Government and obtained the mushroom spawn from Japan and France and started a laboratory in Solan. He started his research experiments on growing edible mushrooms of *Agaricus* and other species, in laboratory conditions.

The Indian Council of Agricultural Research (ICAR), New Delhi has sanctioned grant for the creation of National Centre for Mushroom Research and Training (NCMRT) during VI plan (1982) with the objective of conducting researches on problems of mushroom production, preservation and utilization and to impart training to scientists, teachers, extension workers and mushroom growers. NCMRT started functioning with effect from 1983.

All India Co-ordinated Mushroom Improvement project (AICMIP) was sanctioned by ICAR during VI plan period (1983).

An ICAR sponsored co-ordinate scheme on mushroom Research was in operation at Solan, Bangalore, Ludhina and New Delhi from 1970. This scheme was later converted in to an All India Co-ordinated Mushroom Improvement Project at NCMRT, Solan in 1983.

During the year 1974, incorporation of molasses and brewer's grain-I synthetic compost was adopted on the recommendations of Hayes (FAO, Mushroom Expert). It resulted 7Kg to 13Kg of mushroom per square meter.

UNDP (1982) was launched a mushroom development project at Solan by the State Department of Horticulture. Success in cultivation of the mushroom in H.P caught the imagination of many research workers and growers to initiate similar activity in different regions. This leads to the establishment of various research units of different centers as well as few mushroom farms in the country.

Mushroom cultivation in Kashmir Valley was introduced by Indian Institute of Horticultural Research, Bangalore during 1970. Blue mountain food products have established a big mushroom form at the site which was earlier used by Stewart for mushroom growing. It is cultivated successfully in Down valley (UP) and in Pune. Now it was established and adapted in various districts of AP, Bihar, Haryana, HP, Jammu and Kashmir, Karnataka, Kerala, Madhya Pradesh, Maharastra, Meghalya, New Delhi, Orissa, Punjab, Rajasthan, Tamil Nadu, UP and WB.

Today mushroom cultivation has become a huge export oriented industry and large foreign exchange earning business in India. Many Universities and State govt. Departments of Agriculture and private peoples are giving training in growing mushrooms, which are mostly, exported and also used in a variety of culinary delights in hotels.

Cultivation of white button mushroom

- In India, Bose (1921) was successfully cultured two *Agaricus* mushroom on a sterilized dung media.
- In 1941, Padwick reported successful cultivation of *A. bisporus* from various countries.
- In 1961, a scheme called 'Development of Mushroom Cultivation in Himachal Pradesh' was strated at Solan by the State Govt.in collaboration with ICAR New Delhi. This was the first attempt on the cultivation of *A. bisporus* in India. The technique for its cultivation was evolved using horse dung and wheat straw synthetic compost.
- A modern spawn laboratory and air conditioned cropping rooms were designed and constructed under the guidance of E.F.K. Mental, FAD mushroom expert (1965).
- Numbers of researches were made on evaluation of different strains of mushroom and use of various agricultural wastes and organic manures and fertilizers for preparing synthetic compost.

Oyster mushroom cultivation

Flack (1917) described the first successful cultivation of the oyster mushroom, *Pleurotus ostreatus* on tree stumps and logs. In 1951, Lobwang was the first to grow *Pleurotus* on sawdust mixtures. In 1951, Black, Tsao and Hau grew larger number of fruit bodies on sterile sawdust and oat meal mixture. *Pleurotus flabellatus* cultivation was standardized first in India by Bano and Srivastava during 1962. In 1974, Jandail and Kapoor first demonstrated the cultivation of another oyster mushroom, *Pleurotus sajor-caju* on banana pseudo stem and chopped paddy. Successful cultivation of *P. sajor-caju* is used with paddy straw, sawdust and wood shavings as the substrates.

Paddy straw mushroom cultivation

In India, Su and Seth (1940) have first cultivated *Volvariella diplasia* but the scientific cultivation of this mushroom using spawn was successfully done by Thomas during 1943. In 1939, Department of Agriculture, Tamil Nadu was the first to start the experimental cultivation of paddy straw mushroom. Asthana (1947) reported that addition of red gram dhal to the mushroom beds had increased the yield.

Chapter 4

MUSHROOM RESEARCH STATIONS IN INDIA

1. NCMRT, Chambaghat, Solan- 173 213, Himachal Pradesh.
2. Division of Mycology and Plant Pathology ICAR, New Delhi 110 012.
3. Division of Plant Pathology, Indian Indtitute of Horticulture Research, Hassaragatta 560089, Karnataka.
4. ICAR Research Complex for North Eastern Region, Shilling.
5. Department of Mycology and Plant Pathology, UHF, Solan, HP.
6. Punjab Agricultural University, Ludhiana, Punjab.
7. Department of Plant Pathology, Tamil Nadu Agricultural University, Coimbatore.
8. Department of Plant Pathology, Indira Gandhi Krishi Vishva Vidhyalaya, Raipur, MP.
9. College of Agriculture, Mahathma Phule, Agricultural University, Pune, Maharastra.
10. G.B. Plant University of Agriculture and Technology, Pantnagar, UP.
11. Bidhan Chandra Krishi Vishva Vidhyalaya, Kalyani, WB.
12. Regional Research Loboratory (CSIR), Srinagar, Kashmir.
13. Department of Agriculture, Govt. of Jammu and Kashmir, Srinagar Kashmir.
14. N.D.University of Agriculture and Technology, Faizabad
15. Maharana Pratap University of Agriculture and Technology, Udaipur
16. Kerala Agricultural University, Thrissur ·
17. ICAR Research Complex NEH Region, Barapani ·
18. Horticulture and Agroforestry Research Programme (ICAR Research Complex for Eastern Region), Ranchi
19. Dr.Y.S.Parmar University of Horticulture & Forestry, Nauni, Solan

In addition to the above, other Universities of Agriculture and Horticulture Departments and Krishi Vigyan Kendras are engaged in basic research in mushroom cultivation and training.

The International Society for Mushroom Science

The International Society for Mushroom Science (ISMS) helps the cultivation of edible (including medicinal) macrofungi. It is non-political and non-profit making organisation. The objectives of ISMS are the dissemination of information on new developments and the science of mushroom cultivation and to stimulate exchange of new ideas between growers and scientists around the world. ISMS sponsors a major international congress on mushrooms every 3-5 years. Other seminars, meetings and workshops are endorsed and supported by ISMS working with national committees.

Chapter 5

SCOPE OF MUSHROOM CULTIVATION

Production

Mushroom cultivation has great scope in China, India and in some of other developing countries because of the cheap and easily available raw materials (agricultural waste) needed for the culture. In 1978, China's edible mushrooms production was only 60,000 tones. In 2006, China's mushroom production was over 14 million tones. Now there are more than 25 million people directly or indirectly engaged in mushroom production. Now China has become a leading mushroom producer in the world followed by India and others. It is hoped that the avocation of mushroom farming will become a very important cottage industrial activity in the integrated rural development programme, which will lead to the economic improvement of not only small farmers but also of landless laborers and other poor communities.

The national annual production of mushrooms is estimated to be around 50,000 tonnes with 85% of this production being of button mushrooms. White button mushrooms are grown all over the world and account for 35-45 % of the total mushroom production. In India, large units with production capacities between 2000 – 3000 tonnes / annum, have been set up mainly as export oriented units in the southern, western and northern regions. India exports the highest quantity of the mushroom produced in the country to USA. Netherlands and China account for 60% of the export of mushrooms. Germany is the largest importer and France and UK are large producers as well as consumers. The demand for fresh mushroom is increasing in the international market while that of preserved or canned mushrooms is decreasing. Marketing problem is experienced in the winter   months (December- February) when more than 75% of the annual production comes in market for sale in limited duration and market area. The quantity of mushrooms exported by India in comparison to the world export is almost negligible. Production of mushrooms, especially of the white button mushrooms, in India has gone up during recent years creating marketing problems. The marginal increase in demand is for fresh mushrooms instead of dried/preserved mushrooms. Fresh mushrooms have very short shelf-life and therefore cannot be transported to long distances without refrigerated transport facility. They are sold in the markets in and around the production areas.

- They are good source of protein (20-35% one dry weight basis
- Mushrooms are free from fat except linoleic acid.
- They are good source of minerals and vitamins.
- Mushrooms have low starch content and ideal for diabetic patients also.
- They are considered as low calorie food with traces of sugar and no cholesterol.

- They have medicinal properties like antibacterial, anticancer, hypolipaemic, hypocholesterimic and antihypertension effect.
- The land resources in the world are limited but mushrooms grow in indoor, independent of sunlight and doesnot require fertile land.
- Cultivation of mushrooms is labour intensive and offer vast employment opportunities in rural areas.
- They have huge potential for export as global market is expanding very fast.
- It can be a good companion crop in vinyards.
- Mushrooms provide additional income for farmers.
- The spent up compost can be utilized for manuring and fertilizing the horticulture crops and vegetable crops.

Chapter 6

BIOLOGY OF MUSHROOMS

Introduction

Mushrooms are a saprophytic fungus. It is commonly grows on damp wood, decomposing organic matters like humus, horse dung etc. mushrooms are cultivated commercially in various parts of India. Though different species of mushrooms are available, all showed similar vegetative and reproductive structures.

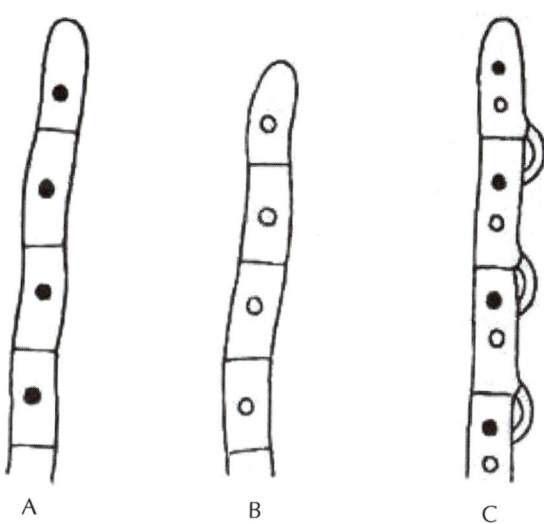

A   B   C

Fig. 6.1 : *A and B Monokaryotic Mycellium; C-Dikaryotic Mycelium.*

Vegetative Body

Vegetative body of mushroom is mycelium. It consists of septate, branched hyphae. It is developed from Spore. Spores on germi-nation develop into monokaryotic or primary mycelium, either + or – type (*Fig 6.1A* and *B*). The primary mycelium is short-lived and it soon transforms into dikaryotic or secondary mycelium by the fusion of two cells of different monokaryotic mycelium (+ and -) following clamp connection (*Fig. 6.1C*).

The hyphae of the dikaryotic mycelia interlace and twist together to form thick white hyphal cord, called rhizomorph which bear the fruit bodies (Button).

Reproduction

Mushroom reproduces by means of vegetative, asexual and sexual.

1. Vegetative Reproduction: In this method, dikaryotic mycelium develops from a spawn. The mass of mycelium divides artificially and produce fruit bodies.

2. Asexual Reproduction: It takes place by chlamydospores. It is formed rarely during unfavourable condition. Terminal or intercalary chlamydospores are developed on dikaryotic mycelium, which on germination produce dikaryotic mycelium. This dikaryotic mycelim develops into fruiting body.

3. Sexual Reproduction: Sex organs are absent in mushroom. Sexual reproduction takes place by somatogamy. A. campestris are heterothallic, but A. brunescens is homothallic. Plasmogamy, karyogamy and meiosis are the methods of somatogamy.

(a) **Plasmogamy:** Two cells of monokaryotic hyphae of opposite strains (+ and -) come in contact with each other (*Fig.6.2*). The cell wall dissolves at the point of contact and a dikaryon (n + n) is formed (*Fig.6.2*). This dikaryo-tic cell develops into dikaryotic myceli-um by regular cell divisions through clamp connection (*Fig.6.2*). The dikaryotic mycelia are subterranean and after aggregation at some points they form button which remains dormant before the rain comes during late sum-mer. After rain, the soil becomes soft and the button develops into fruit body.

(b) **Karyogamy:** It takes place in the young basidium which develops on gills of the fruit body. Both the nuclei fuse together and form diploid nucleus.

(c) **Meiosis:** It takes place soon after Karyogamy and forms four haploid nuclei. The basidiospores, thus formed on the sterigma of basidium are haploid and either of + or – type.

Development of Basidiocarp

The under-ground dikaryotic mycelia aggregate at some points and form a knob-like structure, called button. The button does not grow in dry season and remains hidden one or two inch (2.5-5 cm) below the soil surface. In the late summer with heavy rain, when the soil becomes moist and soft, the button grows rapidly and develops the basi-diocarp (*Fig.6.2*).

During development, the button is differenti-ated into a basal bulbous part and an apical hemispherical region. The bulbous part gradually differentiates into elongated, solid, cylindrical structure, the stipe and the hemispherical region differentiates into a round, convex region, looks like the top of an open umbrella, the pileus (*Fig.6.3*).

Towards the bottom of the hemispherical region some hyphae are drawn apart and form a ring-like cavity, the prelameliar chamber (*Fig. ?*). The upper surface of prelameliar chamber becomes deeply concave and lined with alternating radial bands of slow and rapidly divi-ding cells.

The region with rapid division forms gill-primordia, which develops into gills, that hang downwardly into the prelameliar chamber (*Fig.?*).

The top of the hemispherical region (pileus surface) expands resulting in the increase in radial interspaces between the gills. The edge of the pileus of young basidiocarp connects with the stipe by a membranous tissue called the velum, partial veil or inner veil (*Fig. 4.2E*).

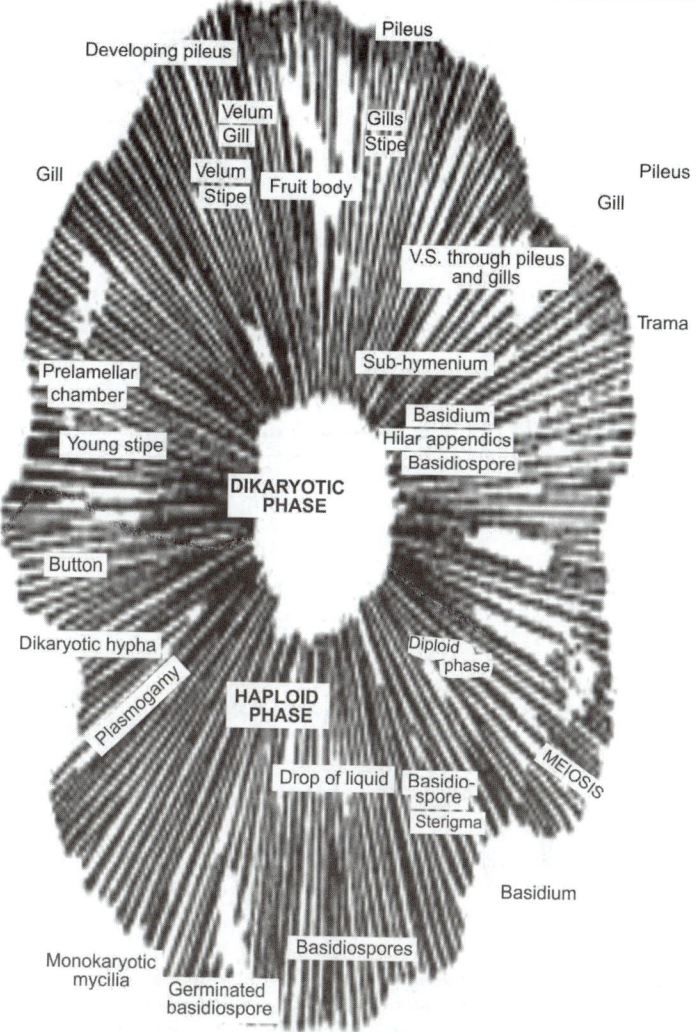

Fig.6.2 : Life cycle of Agaricus.

Further expansion of the pileus causes discontin-uation with velum and the pileus fully expands out like the top of an open umbrella, with nume-rous gills attached on its lower side. The velum remains attached with the upper part of the stipe in the form of a ring, the annular ring or annulus.

Structure of Basidiocarp

A. External Structure: The mature basidiocarp (*Fig.6.4*) is an open umbrella shaped structure with a broad expanded pileus on a long massive stalk-like stipe. The pileus is 5-12 cm (2-5 inch) in diametre with a con-vex upper surface, may be of white, yellow or light brown in colour.

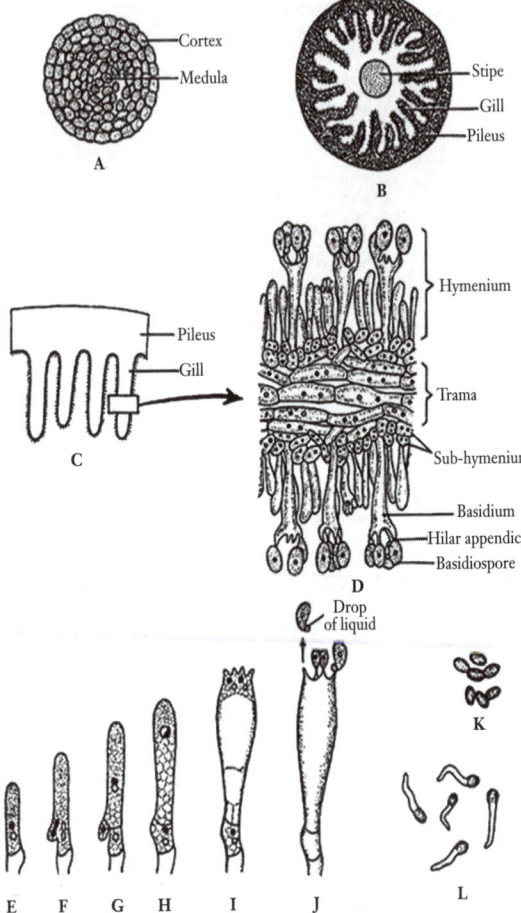

Fig.6.3 : Agaricus. Internal structure of basidiocarp. A Transverse section (T.S.) of stipe, B.T.S. of pileus, C. Vartical section of pileus through gill region, D. Magnified view of a portion of gill. E-J. Development of basidum and formation of basidiospores. K. Basidiospores and L. Germination of basidiospores.

The lower surface of the pileus bears about 300-600 radially arranged gills. The young gills are light pink in colour, but becomes purple or brown at maturity.

The gills never touch the stipe. The stipe is elongated, thick, solid, soft and cylindrical structure and light-pink or white in colour. The fruit body remains attached with the substratum by rhizoidal mycelium.

B. Internal Structure: (a) Stipe: It is differentiated into central medulla, composed of loosely inter-woven hyphae and an outer cortex, made up of densely compacted hyphae (*Fig.6.3A*).

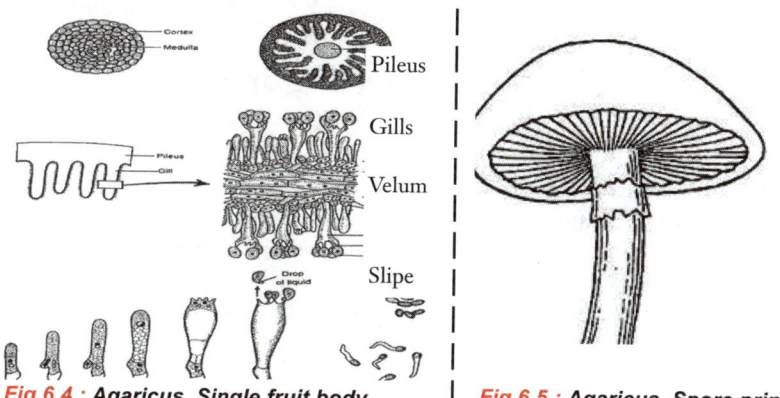

Fig.6.4 : *Agaricus. Single fruit body.* | Fig.6.5 : *Agaricus. Spore print*

(b) Pileus: Like stipe, it is also differentiated into outer compact and inner loose hyphae.

(c) Gills:

It is differentiated into three regions trama, sub-hymenium and hymenium (*Fig.6.3D*):

(i) Trama: It is the central sterile region of the gill, consists of many layers of loosely arranged interwoven hyphae.

(ii) Sub-Hymenium (Hypothecium): It is also a sterile zone, situated on both sides of trama, formed by the late-ral branches of hyphae develops from trama.

(iii) Hymenium: This layer is present on the outer-side of sub-hymenium, composed of fertile basidium and club-shaped sterile paraphyses.

Basidiospore: Basidiospores are oval, thin- walled and uninucleate (*Fig.6.3*).

Dispersal of Basidiospore: The mature basi-diospores are discharged from the basidium by water-drop mechanism. A drop of liquid appears at the hilar appendage. The drop gradually increases in size and creates a pressure which helps the spore to detach from sterigma (*Fig.6.3f*). The spores of a basidium disperse in rapid succession.

Germination of Basidiospore: Falling on suitable substratum, the basidiospore germinates by initiating germ tube which develops into primary mycelium, either + or – strain depends on the strain of spore (+ or -).

Spore Print: It is a technique to study the arrangement of gills in different way. The stipe is removed from the basidiocarp close to the pileus. The pileus is then kept for a few hours on moist cotton. Then the pileus is placed, gills-side down, on a black paper and kept for overnight. Total set up is then covered with bell jar to avoid any foreign distur-bance. The pileus is then carefully lifted and the spores fallen on the black paper become clearly visible. The distribution of spores in a radiating manner indicates a duplicate arrangement of gills. This is called spore print (*Fig.6.5*). This spore may be used for preparation of pure culture.

Chapter 7

CLASSIFICATION OF MUSHROOM

Introduction

Mushrooms are unique in nature. All mushrooms are placed in a division called 'Eumycota'. Eumycota means 'The True Fungi'. The mushroom is the reproductive organ or fruiting body of the fungus. This means that by picking a mushroom we do not harm the fungus. Main body of the organism lies underground in the form of a network of minute threads. Single thread like structure is called 'Hyphae'. These hyphae meet together to form another network called the 'Mycelium'. Mycelium develops in to fruiting body of mushrooms.

Major Classification

Mushrooms belongs to the kingdom fungi. One of the major groups in fungi are Eumycota. Mushrooms belongs to two major subdivisions are (1) Basidiomycotina and (2) Ascomycotina.

Division: Eumycota

Subdividions : (Phylum): (1) Basidiomycotina and (2) Ascomycotina.

I. Basidiomycotina : Mushrooms belonging to Subdivision Basidiomycotina produce their spores 'on' (not 'in') specialized cells called 'Basidia. Sub Division bascidiomycotina have two classes. They are hymenomycetes and gasteromycetes.

1. Hymenomycetes: In this group, the Basidia form an exposed layer called 'Hymenium' on the surface of the fruit bodies. The spores are discharged by force at maturity and thus a spore print is obtainable. It has three orders. They are Agaricales, Aphyllophorales and Tramellales.

(a) Agaricales: The common name for this group is 'The Gilled Mushrooms'. They all have gill like structures on which the Basidia and spores are carried. Eg. *Agaricus bisporus*. It has following families.

   (i) Agaricaceae – *Agaricus bisporus*

   (ii) Amanitaceae - *Amanita muscaria*

   (iii) Bolbitaceae - Agrocybe - *Agrocybe parasitica*

   (iv) Boletaceae - *Boletus sp.*

   (v) Coprinaceae - *Coprinus atramentarius, Coprinus comatus (Lawyers wig), Panaeolus rickenii*

   (vi) Cortinariaceae - *Cortinarius alboviolaceus, Gallerina mycenopsis, Inocybe calamistrata* (Scaly Inocybe)

   (vii) Entolomataceae

   (vii) Gomphidiaceae

   (viii) Hygrophoraceae - *Hygrocybe strangulata, Hygrophorous viridis*

   (ix) Lepiotaceae

Mushroom classification

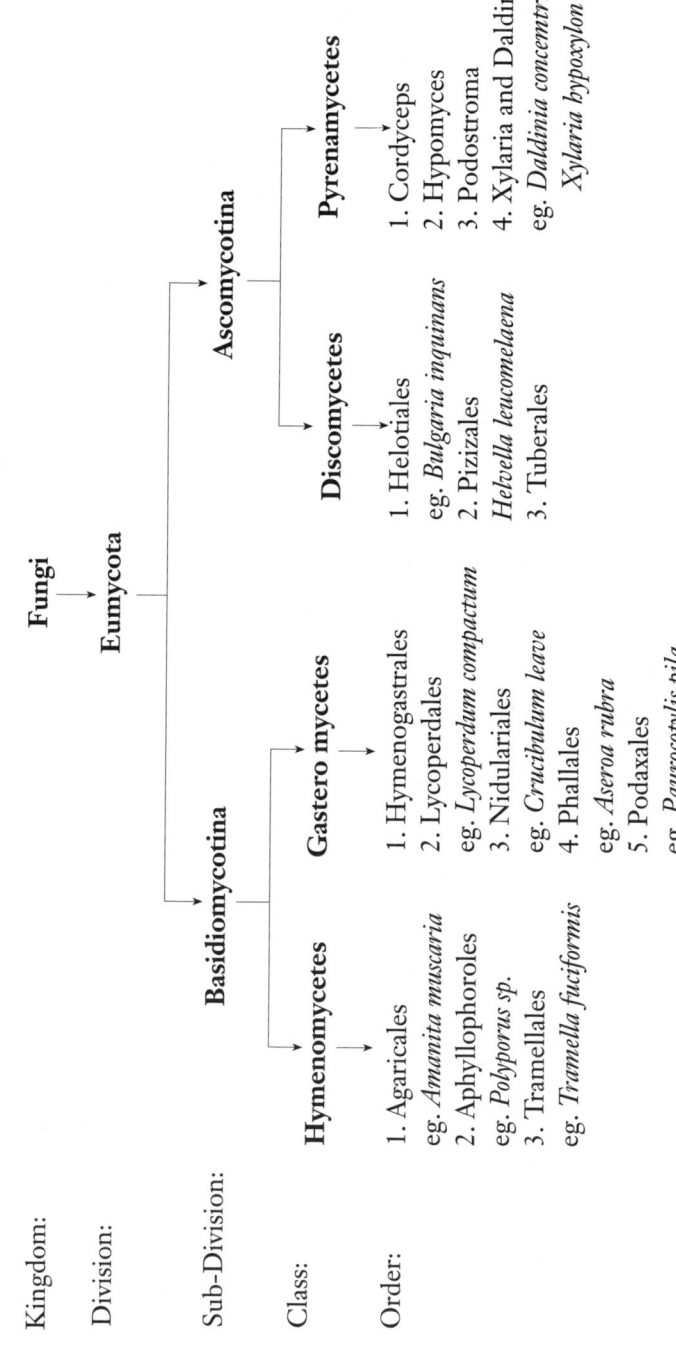

Kingdom: **Fungi**

Division: **Eumycota**

Sub-Division: **Basidiomycotina** | **Ascomycotina**

Class: **Hymenomycetes** | **Gastero mycetes** | **Discomycetes** | **Pyrenamycetes**

Order:

Hymenomycetes
1. Agaricales
   eg. *Amanita muscaria*
2. Aphyllophoroles
   eg. *Polyporus sp.*
3. Tramellales
   eg. *Tramella fuciformis*

Gastero mycetes
1. Hymenogastrales
2. Lycoperdales
   eg. *Lycoperdum compactum*
3. Nidulariales
   eg. *Crucibulum leave*
4. Phallales
   eg. *Aseroa rubra*
5. Podaxales
   eg. *Paurocotylis pila*
6. Tulostromatales
   eg. *Calostoma rodways*

Discomycetes
1. Helotiales
   eg. *Bulgaria inquinans*
2. Pizizales
   Helvella leucomelaena
3. Tuberales

Pyrenamycetes
1. Cordyceps
2. Hypomyces
3. Podostroma
4. Xylaria and Daldinia
   eg. *Daldinia concemtrica*
   Xylaria hypoxylon

(x) Paxillaceae

(xi) Pluteaceae - *Pluteaus umbrosis, Pluteus velutinornatus ; Vovariella speciosa*

(xii) Russulaceae - *Lactarius fragilis* (Candy Cap) , *Russula* sp.

(xiii) Strophariaceae - *Hypholoma brunneum*

(xiv) Tricholomataceae

(xv) Polyporaceae – *Lentinus sajorcaju* (Syn. Pleurotus)

(xvi) Marasmiaceae – *Lentinula* (Shiitake mushroom)

(b) Aphylophorales: This group lacks the spore bearing gills (Except few). They also lack the veil. The spore bearing surface 'Hymenium' can be smooth, wrinkled, bearing teeth like structures (spine like) or a layer of tubes closely packed together.

(i) Cantharellaceae

(ii) Clavariaceae - *Ramariopsis pulchella*

(iii) Ganodermataceae : *Ganoderma sp., Phellinus igniarius*

(iv) Hydnaceae (Teeth fungi): *Hericium clathroides*   (Icicle fungus), *Steccherinum rawakense, Physalaria inflata*

(v) Shizophyllaceae: *Schizophyllum commune*

(vi) Stereaceae (Crust fungi):*Hyphodontia sambuci*

(vii) Dacrymycetales - Dacrymycetaceae : *Calocera fusca, Dacryopinax spathularia*

(c) Tremellaceae (Jelly fungi) : The fruit body is an irregular shape and usually found in the bark of dead branches. It is up to 7.5 cm broad and 2.5 to 5.0 cm high, rounded to variously lobed or brain-like in appearance. The fruit body is gelatin-like but tough when wet, and hard when dry. The surface is usually smooth, the lobes translucent, deep yellow or bright yellow-orange, fading to pale yellow, rarely unpigmented and white or colorless. The fruit bodies dry to a dark reddish or orange. The spores, viewed in mass, are whitish or pale yellow.

Auricularia lacteal, Auricularia polytrica, Pseudohydnum gelatinosum, Tremella fuciformis.

2. Gasteromycetes : The spores are produced inside a spore case therefore never has been exposed to external conditions. It contains six orders.

(a) **Hymenogastrales**: The Hymenogastraceae is a family of fungi. The blue-staining members of the genus Psilocybe form a clade that is sister to Galerina in the Hymenogastraceae.

(b) **Lycoperdales**: This group of mushroom has no cap, gills or stems. It has round shaped 'Spore Case'. The spores inside the ball are called the 'Spore Mass'. At maturity, the ball ruptures or develops an opening at the top through which the spore mass. e.g., Lycoperdum compactum

(c) **Nidulariales** (Bird's nest fungi): This family members look like bird nests with the eggs. These are usually flat like capsicum seeds and carry the spores within them. They can be found growing on the ground, dung or wood. Usually minute in size. e.g., *Crucibulum laeve*.

(d) **Phallales** (Stink horns): This family contains different forms of mushrooms. They start off looking like a puffball ('Peridium').

(e) **Podoxales** (Pouch fungi): This group resembles the *Agarics* in the presence of a cap and a stem, but differ in that the spores are not forcibly discharged at maturity but rather they are kept enclosed within the cap until the mushroom deteriorates or is damaged by animals or insects for dispersal. E.g. *Paurocotylis pila*. Cap is pouch like. Orange to red in colour.

(f) **Tulostomatales:** Tulostomatales is an order of gasterocarpic basidiomycetes related to Agaricales. Common name: stalked puffballs. *Podaxis pistillaris.*

Table 1 : Classification of Mushroom

Kingdom	Division	Subdivision	Class	Order
Fungi	Eumycota	**Basidiomycotina**	Hymenomycetes	Agaricales
				Aphyllophorales
				Tramellales
			Gasteromycetes	Hymenogastrales
				Lycoperdales
				Nidulariales
				Phallales
				Podaxales
				Tulostomatales
		Ascomycotina	Discomycetes	Helotiales
				Pezizales
				Tuberales
			Pyrenomycetes	Cordyceps
				Hypomyces
				Podostroma
				Xylaria and Dalinia

Subdivision : Ascomycotina

Ascomycetes are mushrooms that produce their spores inside a sac like cells called 'Asci, (Ascus). There are two classes, they are discomycetes and Pyremomycetes.

1. Discomycetes: These are ascomycetes in which the hymenium, (spore carrying surface) lines an area on the fruiting body exposed to the elements. This is similar to the basidia of the Agaricus. It has three orders.

(a) **Helotiales** (Earth Tongues): These mushrooms have either stalked or cup like apothesia. *Helotiales* is the largest order of inoperculate discomycetes. It contains the famous blue-green cup fungi. It grows mostly on oaks. Its asci are only slightly thickened in contrast to other Leotiomycetes. Most *Helotiales* live as saprobes on soil humus, dead logs, manure, and other organic matter.

(i) Bulgaria inquinans ; (ii) Ciboria -Bisporella citrina; (iii) Geoglossum cookeianum ; Trichoglossum hirsutum ; (iv) Leotia lubrica; v) Microglossum and Spathularia

(b) **Pezizales:** The order contains approximately 1125 species organized in 16 families. It contains a number of species of economic importance, such as morels, the black and white truffles, and the desert truffles. The Pezizales are saprobic, mycorrhizal, or parasitic on plants. Species grow on soil, wood, leaves and dung. Most species occur in temperate regions or at high elevation. Mushrooms in this group are characterized by that their asci have lids which open when the fruit body is disturbed and through this it releases the spores.

(i) Helvellaceae - *Helvella leucomelaena* (White-Footed Elfin Cup)

(i) Pezizaceae - Pithya vulgaris

(ii) Sarcosomataceae - *Plectania campylospora* (Black cup fungus)

(c) **Tuberales** (Truffles): These are rare mushrooms. They are rarely seen because they grow underground like plant tubers.

2. Pyrenomycetes (Flask Fungi) : In this group, the spore bearing surface is carried within the fruit body. It has following orders.

(a) Cordyceps

(b) Hypomyces

(c) Podostroma

(d) Xylaria and Daldinia.

Chapter 8

EDIBLE MUSHROOMS

Most species of mushrooms are edible and only few are poisonous. Non edible or poisonous mushrooms are either very dangerous or do not possess any aroma at all. There is no confirmed method to distinguish poisonous mushroom from an edible mushroom.

Edible mushrooms

Edible mushrooms include thousands types of mushrooms that are harvested from wild and others that are not easily cultivated. *Agaricus bisporus*, is generally considered safe for most people to eat because it is grown in controlled environments. Several varieties of *A. bisporus* are grown commercially, including whites, crimini, and portabello. Other cultivated species now available at many grocers include Shitake, maitake or hen-of the wood, oyster, and enoli.

Some species are difficult to cultivate particularly mycorrhizal species. Some of these species are harvested from the wild, and can be found in markets. When in season they can be purchased fresh, and many species are sold dried as well. The following species are commonly harvested from the wild:

- *Boleus edulis* or edible Boletus, native to Europe. It is also called Pig mushroom in italy, in German as Stone mushroom, in Russia as "white mushroom", and in French the *cep*. It is also known as the king bolete, for its delicious flavour. It is sought after worldwide, and can be found in a variety of culinary dishes.

- *Cantharellus cibarius* (chanterelle) is one of the best and most easily recognizable mushrooms, and can be found in Asia, Europe, North America and Australia.

- *Grifola frondosa*, known in Japan as *maitake*. It is also called as"hen of the woods" or "sheep's head". It is a large, hearty mushroom commonly found on or near stumps and bases of oak trees, and believed to have medicinal properties.

- *Gyromitra esculenta* is also called "false morel". It is a high prized mushroom by the Finns. This mushroom is deadly poisonous if eaten raw.

- *Hericium erinaoeus*, a tooth fungus; also called "lion's mane mushroom."

- *Hydnum repandum*, a sweet tooth fungus, hedgehog mushroom, urchin of the woods.

- *Lactarius deliciosus* is otherwise called 'Saffron milk cap' – a valuable mushroom in Russia.

- *Morchella* belong to the ascomycete grouping of fungi. They are usually found in open scrub, woodland or open ground in late spring. When collecting this fungus, care must be taken to distinguish it from the poisonous false morels, including *Gyromitra esculenta*.

Other edible wild species

1. *Amanita caesarea* (*Caesar's Mushroom*)
2. *Armillaria mellea*
3. *Boletus badius*
4. *B. elegans*
5. *Chroogomphus rutilus* (*pine-spikes or spike-caps*)
6. *Calvatia gigantea* (*Giant Puffball*)
7. *Clavariaceae species* (*coral fungus family*)
8. *Coprinus comatus*
9. *Hygrophorus chrysodon*
10. *Lactarius salmonicolor, L. subdulcis* (*mild milkcap*) and *L. volemus*
11. *Laetiporous sulphureus* (*Sulphur shelf*)
12. *Leccinum aurantiacum* (*Red-capped scaber stalk*) & *L. soabrum* (*Birch bolete*)
13. *Polyporus sulphureus*
14. *Ramariaceae species* (*coral fungus family*)
15. *Russula sp.* and
16. *Sparassis crispa* (*cauliflower mushroom*).

Chapter 9

IDENTIFICATION OF EDIBLE MUSHROOMS

There are many traditional methods for testing mushrooms but they are unreliable or incorrect. These are the some false believes about the poisonous mushrooms.

(a) Mushrooms are growing in the meadow are edible and those growing near serpent holes or on trees are not edible.

(b) Coloured mushrooms are poisonous and white or grey ones are edible (*Cantharellus cibarius* and *Tricholoma nudans* are bright coloured but quite edible. But *Amanita phalloides*, *A. verna* and *A. verosa* are completely white and deadly poisonous).

(c) Edible mushrooms rind off easily and they do not change silver spoon or garlic black when boiled while cooking. This method of testing is not correct as the poisonous mushroom, *A. phalloides* did not change silver spoon black.

(d) Cap of mushroom is eaten by a snail or an animal can be safely eaten by human beings. This is not correct; rabbits eat poisonous mushrooms which are neutralized in their stomach.

(e) Edibility based on colour, shape, smell and tastes, exudation of milk are also found incorrect. *Eg. Agaricus xanthoderma* change the colour from white to yellow when it was touched or rubbed, later it was changed from yellow to chocolate.

(f) In the same genus one may be poisonous and the other one is edible. *Eg. Lepiota margani* is a poisonous one but *L. rachodes* is edible.

People in rural areas who know much about appearance of mushrooms in different seasons believed that the mushrooms appearing in spring are poisonous and those which appear during autumn are edible. Practically this is not correct, both poisonous and edible mushrooms came up during autumn and when they are eaten may cause death.

A mushroom species may edible at young and fresh, but it may become poisonous, when it over mature or decays. Edible mushrooms may cause harmful effects when cooked or consumed with alcohols. *Eg*, the mushroom Ink Cap *Coprinus atromentarius* causes skin discolouration when it was taken along with alcohol.

Many fungi cause a reaction with in 30 to 60 minutes. Some deadly poisonous mushrooms may not produce noticeable symptoms until 8 to 12 hours. Some species lose their toxicity on drying, few fungi especially those containing amatoxins are poisonous.

The best way is to identify the mushroom by means of well illustrated manuals. The following precautions are useful to avoid mushroom poisoning.

1. Mushroom should not be collected at button or immature stage.

2. Edibility test should be started with a small piece of mushroom.

3. Some mushrooms will ooze out white milky juice when they are cut. Such mushrooms are to be avoided.

4. Over mature or rotten and bad odour emitting mushrooms should not be eaten.

5. Avoid eating of mushroom with alcohol.

Summary

Fruit bodies are grey in colour.

No valva cup.

Have pleasant taste

Grow in dead organic matter.

Grow on open sunny environment.

No gelatinous surface.

No fibrous pileus.

Chapter 10

POISONOUS MUSHROOMS

There are a number of species of mushroom that are poisonous. More generally, and particularly with gilled mushrooms, separating edible from poisonous species requires meticulous attention to detail. There is no single trait by which all toxic mushrooms can be identified, nor one by which all edible mushrooms can be identified. Some resembles edible species, eating them could be fatal. However even *A. bisporus* contains 'agaritine' which metabolizes when eaten into hydrazine, which is carcinogenic, but this chemical is largely or completely removed by cooking.

In general the chemical properties of mushrooms are the fact that many species produce secondary metabolites that render them toxic, mind-altering, or even bioluminescent. Toxicity likely plays a role in protecting the function of certain mushrooms. The evolution of chemicals render the mushroom is inedible, either cause vomiting, or allergic to consumption.

Poisonous mushrooms may be identified on the basis of some morphological characters. But this is not exactly accurate.

1. Mushrooms having ring on its stalk are generally poisonous (*Amanita virosee*) except some edible mushrooms like *Agaricus compestris*.

2. White capped mushrooms should be avoided.

3. Fresh mushrooms with pores are not suitable for consumption.

4. Mushrooms having swollen stalks are poisonous.

5. Little brown mushrooms are generally toxic and its consumption is not safe.

The method of preparation and the way of consumption also have considerable toxic effect.

1. Some mushrooms are poisonous when eaten raw but it becomes edible after cooking.

2. Some mushrooms are poisonous when consumed at larger quantity.

3. Some mushrooms are poisonous to children

4. The mushroom *Gyromitra esculenta* contain a toxin which is water soluble. This toxin leaches to the water during cooling. After cooking it become edible.

5. Some mushrooms are poisonous when consumed with alcohol.

Some important poisonous mushrooms are listed below,

(a) Many of the species in *Amanita* are poisonous. They include 'Death cap': *Amanita phalloides* – which produces amanitin and phalloidin toxins (***Fig.10.1***). 'Fool's mushroom' – *A. verna*, Angel mushroom – *A. verosa* and the stems of the *A. muscaria* are predominantly poisonous. *A. phalloids* and *A. verna* are highly toxic.

Fig.10.01: Poisonous mushrooms - Macrolepiota and Amanita partherina

(b) *A. muscuria* is also a poisonous species which results in death of healthy person. After intake person suffers from vomiting and diarrhoea, loss of memory and have tendency to sleep. The active toxic principle in it is mycetomuscarine and muscardine.

(c) *Inocybe patouidare* can be identified by the presence of conical cap, absence of ring on the stipe and inrolled margin. This mushroom becomes brownish red during handling. The poison cause diarrhea, loss of memory and tendency to sleep.

(d) *Lepiota margani* and *Agaricus xanthoderma* are important poisonous mushrooms.

(e) *Gyromitra esculenta* causes gastric entritis.

(f) *Amanita rubescens* causes blood destroying activity.

(g) *Entoloma lividum* and *E. sinuatum* are poisonous. It causes violet sickness, headahe, stomach pain, vomiting and diarrhoea.

A piece of one cubic centimeter size can cause several illness or even death. Recovery after consumption of these is difficult. All parts of these fungi are dangerous. The death occurs 3 to 10 days after ingestion. The poison causes paralysis of nervous system and degenaration of the liver cells.

Chapter 11

MUSHROOM POISONING

1. Nerve poisoning: this type of poisoning occurs 2-3 hours after eating the mushrooms. The consumption of *Amanita muscuria*, *A. pantherina* and *Inocybe pathotillardii* causes this type of poisoning. The toxic effects are due to muscarine and ibotenic acid contained in these mushroom. The fatal death rates are 10-15%.

2. The cellular poisoning: *Amanita* species and *Galerina* species are mainly responsible for this type of poison. Amanita toxin attacks body tissues and cause death. Small quantity of this mushroom (~10mg) enough to kill healthy human. This toxin is thermostable and it is not removed by boiling. This toxin inhibits the intra cellular functions such as gene expression and protein synthesis. The symptoms of this toxin causes hepatic renal failures.

3. Digestive poisoning: Mushrooms such as *Poletus satanas*, *Emtomola lividum*, *Lactarcius fornosus* and *Rossula* sp. cause this type of poisoning. Toxicity is extremely variable from individual to individual mushrooms. Norcaperatic acid has been found to be the common cause of this poisoning. Symptoms such as nausea, vomiting and diarrhoea occur rapidly.

4. Muscular poisoning: the muscular disorders appear after 30-60 minutes of consumption of *Psilocybe cubensis*. The symptoms include excitement in muscular system especially in the smooth muscle fibres and the patient may feel muscle weakness and drowsiness. This type of disorder appears due to psilocybin toxin.

5. Poisoning effect of mushroom with alcohol: This type of poisoning is reported if Coprinus atramontaminus commonly known as Ink cap is consumed with or after drinking alcohol. The symptoms occur in short time. Flushing of the face and neck with aching tightness of neck veins, a feeling of swelling parenthesis in the hand and feet followed by chest pain. Later nausea, vomiting, sweating, visual disturbances, weakness, confusion and respiratory difficulties may occur.

The following first-aid measures should be provided to the affected persons,

1. To remove the eaten part of mushroom from the system.

2. The toxin absorbed by blood should be exhausted or eliminated.

3. Pay prompt attention on the condition of patient specially collapse and generally to keep a watch on the action of heart.

Some preliminary treatments are as follows :

1. Expulsion of the fungus: if the poisonous mushrooms are consumed an attempt should be made for vomiting as soon as possible. If vomiting is possible in normal condition give a teaspoon of mustard oil in half glass of water. A pomorphin and zinc sulphate may also used under medical direction. If possible stomach should be washed out by means of a stomach tube. The next step is 1

or 2 spoon of sulphate of soda or sulphate of magnesium in a glass of warm water, followed by a table spoon of castor oil in a little milk and two table spoon of olive oil with egg white yolk.

2. Elimination of toxin: the poison already absorbed in the blood may be eliminated or exhausted by subcutaneous injection of atropine and by other medical means.

3. Depression, sleepiness, leatheriness can be treated by the use of strong tea or coffee etc.

4. Prolonged and excessive vomiting can be checked with soda water or by giving patient a small piece ice to suck.

Chapter 12

PURE CULTURE PREPARATION OF MUSHROOM

Introduction

Mushroom is an interesting modification of fungal mycelium. They are non-green fungal plants. There are more than 2000 edible species of which only a few have been brought under cultivation on commercial scale. The species grown more commonly and having good export potential are, *Agaricus bisporus* (white button mushroom), *Volvariella spp.* (paddy straw mushroom), *and Pleurotus spp.* (Oyster mushroom). Spawn of mushroom is called seed of plants. Spawn needs pure and mother culture.

Pure culture preparation

There are two ways of raising pure culture. They are tissue culture and spore culture.

Base Spawn / Nucleus Culture/ tissue culture

In tissue culture a well grown mushroom with membrane covering the gills is selected and from which a small bit of mushroom from gill portion is taken using forceps and inoculated on PDA or MEA media slants under aseptic condition (PDA – potato dextrose agar, MEA- Malt extract agar are the culture media readily available in the market). The mycelium covers the entire surface in a week's time and culture becomes ready for further multiplication.

Procedure (*Fig.10.1*)

Select well grown, disease free button mushroom early in the morning and keep it on a clean paper for 2-3 hr, to get certain amount of moisture present in the mushroom to get evaporated.

Clean the culture room/ laminar flow chamber with antiseptic solution.

Keep the sterilized PDA slants, razor blades, forceps etc. inside the chamber and put on he UV light.

After 20 minutes put off the UV light and start working after 5 minutes.

Sterilize all the instruments to be used by exposing to Bunsen burner.

Take in the mushroom and split open the mushroom longitudinally into two halves.

Using a blade cut a small piece of tissue from the centre of the spilt mushroom at the junction of pileus and stipe.

Remove the cotton plug of the agar slant and the tissue is aseptically placed inside the slant by using a sterilized forceps and closes it immediately.

After transferring tissues from the mushroom , the tubes are arranged in a wire basket and kept in a clean room at room temperature for the growth of the fungus.

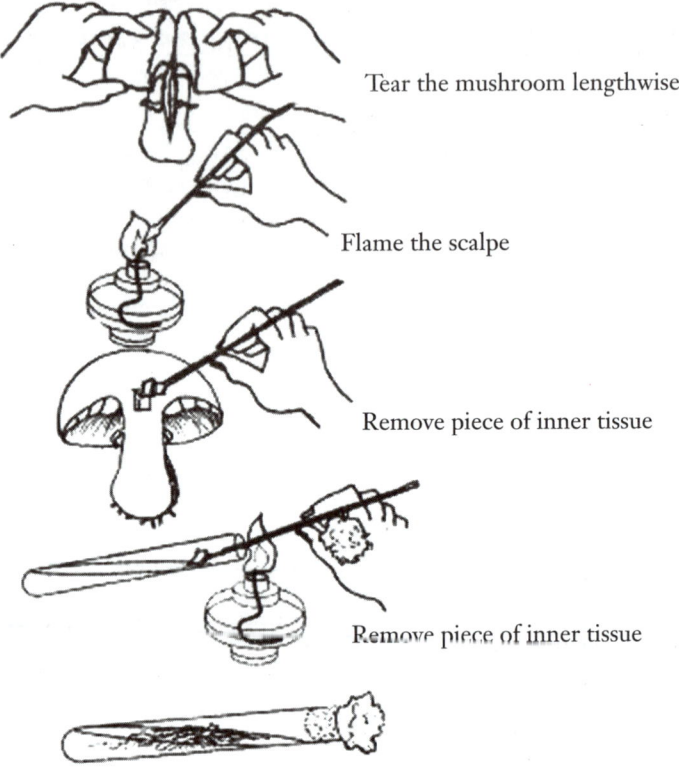

Tear the mushroom lengthwise

Flame the scalpe

Remove piece of inner tissue

Remove piece of inner tissue

Within a few days mycelium will branch out on the agar

Fig.10.1 : **Procedure.**

Observe the tube at periodical intervals and remove the contaminated ones. The tubes will be ready for further use within another ten days. The base spawn is used for preparation of mother spawns.

Spore culture

In spore culture method, the spores are collected from well developed fruiting body by 'spore mapping technique' and then the spores are inoculated to the PDA or MEA slants as in tissue culture under aseptic condition.

Multi-spore culture is made from spore print that can be obtained by hanging a alcohol sterilized fresh fruit body on a loop of wire above a petriplate/ sterilized paper. Spores are serially diluted and transferred to sterile potato-dextrose-agar (PDA) or malt-extract-agar (MEA) culture slants. These slants are then incubated at $25°C \pm 2°C$ for 2 weeks to obtain pure culture.

Mother Spawn: Mother spawn is nothing but the mushroom fungus grown on a grain based medium. Sorghum grains are the best substrate for excellent growth of the fungus. Well-filled, disease- free sorghum grains are used as substrate for growing the spawn materials. The various steps involving in preparation of mother spawn are listed below here under.

Procedure

Wash the sorghum grains in water thoroughly to remove chaffy and damaged grains.

Cook the grains in an autoclave / vessel for 30 minutes just to soften them.

Take out the cooked grains and spread evenly over a Hessian cloth on a platform to remove the excess water.

Mix Calcium carbonate ($CaCO_3$) thoroughly with the cooked, dried grains @20 g/kg.

Fill the grains in polypropylene bags up to ʃ th height (approximately 300-330 g/bag), insert a PVC ring , bold the edges of the bag down and plug the mouth tightly with non-absorbent cotton wool.

Cover the cotton plug with a piece of waste paper and tie tightly around the neck with a jute thread.

Arrange the bags inside an autoclave and sterilize under 20 lbs. pressure for 2 hours.

Take out the bags after cooling and keep them inside the culture room and put on the UV light.

After 20 minutes put off the UV light and start working in the culture room. Cut the fungal culture into two equal halves using a inoculation needle and transfer one half portion to a bag. Similarly, transfer another half portion of the culture to an another bag.

Incubate the inoculated bags in a clean room under room temperature for 10 days for further use to prepare bed spawn.

Culture maintenance

Good mushroom fungal culture is the nucleus for successful production of any mushroom under cultivation. In mushroom cultivation, culture collection is very important. Successful operation of mushroom production highly depends on the proper maintenance of pure cultures of mycelium capable of producing fruiting bodies of high productivity couples with excellent desired characters.

Once a mushroom culture is selected as a vigorous, desirable and high productive strain, it should be preserved as stock culture in a mycelium bank, it is an essential component in a mushroom industry. Normally the cultures isolated are maintained as slant cultures. When tissue cultures or isolation of spores, the resulting culture should first be tested for yield and other characters.

Potato dextrose agar is prepared as per the standard procedure. Inoculate the pure culture of mushroom spores on the media and incubate it for 2 days. Then the test tube culture (slant) and plates are stored at 4°C for 25–30 days. Repeated sub culturing is required for maintaining active mushroom spores.

Storage of mycelium for longer period

Mushroom cultures are normally labeled and maintained in a suitable agar medium in glass test tubes. *Volvariella* cultures are normally maintained in agar medium and stored at 15 to 20°C. It is stored in refrigerator as the fungus does not survive below 10°C. Cultures stored at room temperature are simply transferred to new agar media at least for every 3 to 5 months or when the agar slants drying up. In this method mite contamination is possible for which sealing the tubes with cigarette papers or aluminium foil is advocated. To prevent mites as well as drying of agar add previously sterilized mineral oil to the fully grown fungal culture. Mineral oil is added up to a depth of 1cm above the top of the agar slant of fully grown cultures. If oil is to be added, screw capped test tubes should be used instead of test tubes with cotton plugs. Viability of the culture should be tested after two years of storage. Retrieval is done by cutting a mycelial plug from the stock culture. The same procedure is used for further sub culturing.

Propagation and multiplication of mushroom mycelium

Spore cultures or tissue culture spores are used as the parent culture. In tissue culture some of the properties are different from the original strain. Hence it is safe to test these new cultures with the original or parent strain, before keeping them as stock cultures or using them for making spawns.

After selection of a culture, it has to be tested for propagation and cultivation. Then the culture is to be transferred to many test tubes containing the selected medium, viz., PDA for stock. 10 to 20 numbers of test tube cultures are ideal for sub culturing. The culture is allowed to grow at optimum condition and is sealed with paraffin or screws capped or aseptically added with mineral oil. Then it is stored in refrigerator (5 to 10°C) for long storage.

Originally isolated strain is called as F1 generation, due to subsequent multiplication and sub culturing in test tubes or flat bottle cultures called F2 generation. At this stage these test tube or flat bottle cultures are known as subcultures. Then it is transfer to the spawn bottles, the cultures in the spawn bottle (F3) is called mother spawn. Take the mother spawn bottle and transfer its mycelium to 40 or more spawn bottles. The mycelium grows on the sorghum grain (substrate) and the culture in this bottle (F4) is called as plating or bed or commercial spawn. This is used for growing of mushroom from one stock culture minimum of 2500–4000 commercial or plating spawn bottles of F4 quantity can be obtained. While the subculture (stock) prepared from the F1 stock culture is called F2 stock, the commercial spawn culture is produced in F5 generation. When the spawn have reached F7 to F8 quality, it is necessary to isolate again from the mushroom to start F1 stocks.

Chapter 13

STRAIN DEVELOPMENT

The reservoir of edible mushrooms, like other microorganisms used for industrial purposes, is not unlimited. It is generally recognized that in order to maintain and breed high-yielding strains, the techniques employed in mushroom breeding should now and then be modified and improved in accordance with new findings and progress in the scientific world as a whole, and in microbiology and genetics in particular. Collection, Introduction & Selection, Anastomsis (Strain mixing), Hybribdization, protoplast fusion are some of the stain improvement methods

1. By Collection, Introduction and Selection : Collection of mushroom culture obtained from nature in different countries throughout the worl are important for future strain improvement. This type of strain is called exotic strain. These strains have been evaluated time to time , under various growing conditions. An introduction and selection has been carriedout in edible mushrooms for evaluating superior variety among introduced strain. Selection is done by monosporic culture in homothallic species like volvariella and Agaricus. However for heteriothalic species tissue culture method is suitable. Single spore culture is suitable for fast growing strains

2. By hybridisation : It is one of the most suitable methods for cross breeding. In addition to the conventional method of mating between two genetically compatible strains through which dikaryon mycelia and fruiting bodies are formed, steps towards a broader spectrum of hybridisation can be achieved in strain improvement of edible mushrooms by the following ways.

(a) **Use of auxotrophs.** Auxotrophs can be obtained naturally or induced by mutagenesis. The contrasting auxotrophs can be paired and the products can be screened for hybridity on minimal medium. Certainly, the feasibility of auxotrophs to be used as a tool for hybridization depends on how easily auxotrophs can be obtained in the strains of the mushroom.

(b) **Use of resistance markers.** Mutants resistant to antimetabolites have been suggested recently as alternatives to auxotrophs for use in mushroom breeding programmes. The treated spores or hyphal fragments, which can be grown on a medium containing an inhibitory concentration of the anti-metabolite, would be considered to possess the marker. Complementary resistant strains would be grown together and, hybridity can be confirmed by transferring it to a medium containing the two appropriate antimetabolites.

(c) **Protoplast fusion** : One of the most effective barriers to sexual reproduction is the inability of hyphae from two selected strains to fuse. Several laboratories have reported that protoplasts can be isolated from plant and microbial cells by enzymatic breaking of the

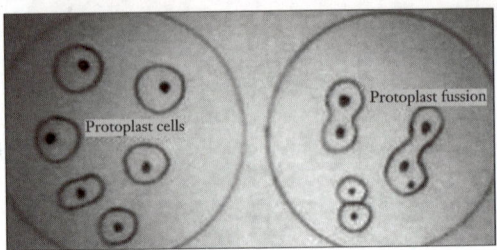

Fig.13.1 : Protoplast Fusion.

cell wall, in the presence of an osmotic stabiliser. Such protoplasts can be effectively induced to fuse in the presence of polyethylene glycol (PEG). After a short period of time protoplasts can regenerate their cell walls and start to propagate as normal cells or hyphae. These cells are heterokaryotic if fusion occurred between cells from genetically different strains. This can serve to increase the frequency of intraspecies crosses in organisms in which natural mating rarely occurs. The technique has even a much wider application, and can be used for interspecies and intergeneric crosses in some organisms, which normally cannot be crossed. Although such an approach has been carried out in several laboratories, until now, no clear and economically applicable results have been reported in edible mushrooms (*Fig.13.1*).

(d) **Anastomosis:** This will occur when two strains are grown in a single environment.

Chapter 14

MUSHROOM FARM LAYOUT

Properly equipped farm is needed for effective low cost mushroom cultivation. Main aim of mushroom farm is be able to achieve maximum yield with least possible trouble and expense.

The land selected for mushroom cultivation should possess the following characters.

1. Site should connect to road.

2. Should have 5 meter wide road, with specific space for turing vehicles.

3. Availability of regular water supply.

4. Availability of regular electricity supply.

5. Availability of labour and raw materials.

6. Sight should near to market.

7. Should have sufficient space for future expansion.

8. Availability of proper drainage system.

An ideal mushroom unit should have the following infrastructure (*Fig.14.1*)

1. Compost Yard (A)

2. Bulk Chamber / Pasteurization chamber (B)

3. Spawning area (C)

4. Spawn running room (D1, D2)

5. Casing pasteurization room (E)

6. Boiler room (F)

7. Growing Room (G1-G10)

8. Generator room (H)

9. Air conditioning room (I)

10. Spawn laboratory (J)

11. Post harvest handling unit (K)

Other facilities needed are **Cleaning chamber, Cropping room, Office room, Room for post harvest handling, Store room.**

Construction of infra structure

Composting yard (A)

It is required with or without side walls.

Around 6000Sq. feet (100" × 60" × 20'h) size composting yard is required for the preparation of 25 tones of compost.

It is made with cement floor and proper drain for liquid leaking from manure.

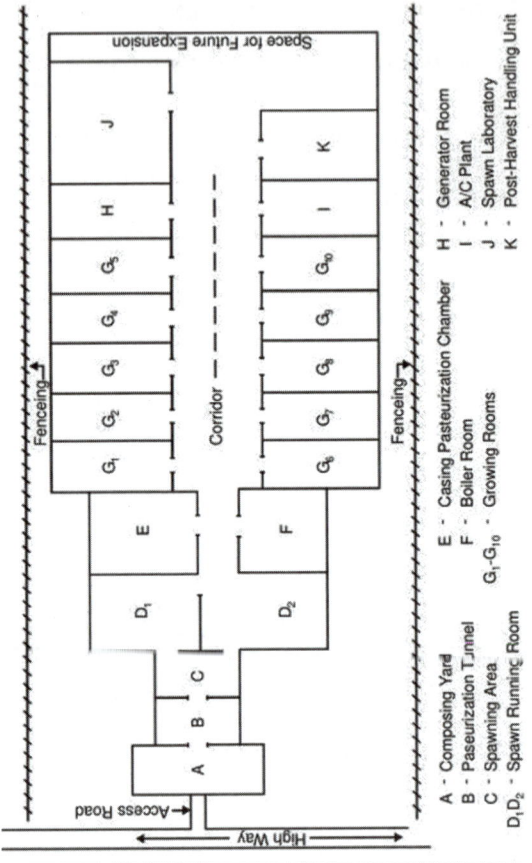

Fig.14.1 : General layout of a mushroom unit.

All drained liquid should be collected in a common tank. This liquid can be sprayed over partial compost.

Yard should be covered with GI roofing to protect the piles against dehydration, rain, snow and wind. Foundation should be 2 feet height. Underground aeration should be done with perforated pipes.

Pasteurization chamber / Tunnel

It is used for phase II of composting. This needs two types of chamber. They are heating chamber and bulk pasteurization tunnel.

Peak heating chamber:

It is also called single zone system. It consists of an insulated room with steam injector and recirculation facilities. Operations like pasteurization, conditioning, spawn running, casing and cropping are done in this chamber.

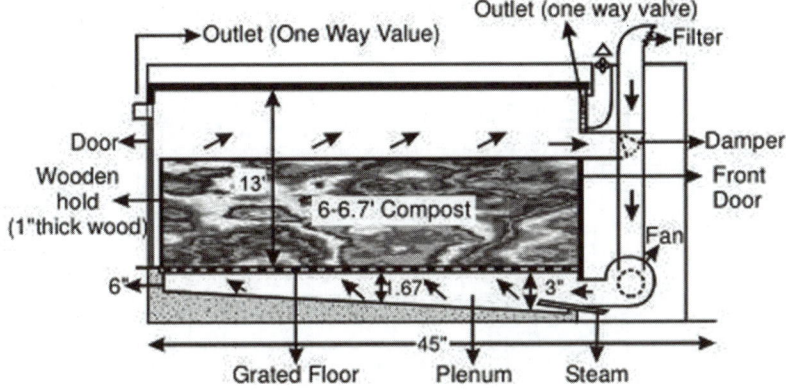

Fig.14.2 : Tunnel of bulk pasteurization.

Bulk pasteurization chamber/ tunnel:

It is also called double zone system. Here compost is filled into specially built chamber. Most commonly used tunnel are of the dimention of 45'x8'x13'h. The thickness of bed may vary depends on straw content, hardness, quality of animal matter, moisture content and degree of composting.

Foundation of bulk chamber should be strong enough preferably on firm base/ ground. A well insulated floor can be obtained by spreading a layer of plastic sheet on a smooth layer of sand of about 8". On the top of this is poured a weak floor of 2" thickness of concrete. After this floor is covered with sheets of materials containing asbestos or thermocol sheets (*Fig.14.2*).

Casing pasteurization chamber (E)

The wall of casing chamber should be made of cavity walls of concrete of 8" thick with an air space of 2" between them. The dimensions of chamber depends on the quantity of casing soil that must be steamed. On the inner walls, asbestos cement sheet will be placed crosswise on support. After this, a layer of 16" thick foam glass or 6" thick rock wool is spread on ceiling. Proper arrangement for drainage should be provided.

Cropping room (G1-G10)

Construction of cropping room is as per our requirement. Two types of crooping rooms may be constructed. They are

1. Natural environmental condition growing room.

2. Environmentally controlled condition growing rooms.

Natural environmental condition growing room.

It is constructed using single brick wall with cement blaster. Roof is made with RCC or asbestos with falce ceiling. Exhaust fan should be used with proper vents. Most of the places thatched hut is used for providing natural environment. Gunny bags may be used for maintaining moisture content (*Fig.14.3*).

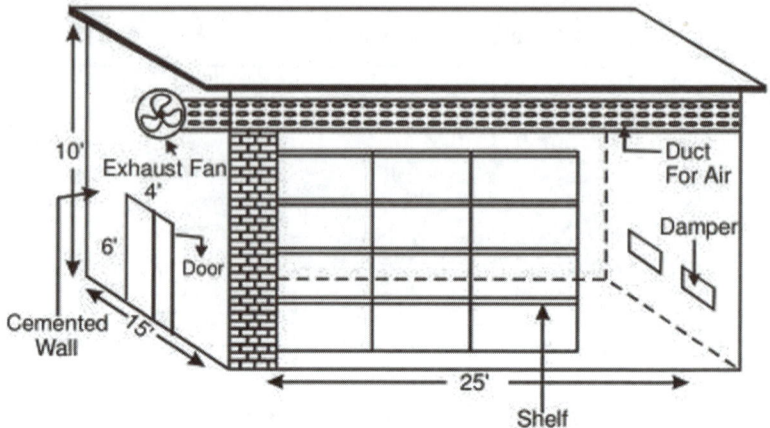

Fig.14.3 : Natural environmental conditions growing room.

Racks may be made with wood or steel.

Size of the room: 60'x30'x10"h for handling 20-25 tonnes of compost.

Environmentally controlled condition growing rooms.

Wall of the cropping room is insulated with 2" thick insulating materials. Floors are made with reinforced concrete with smooth finishing. Sewage system should be well laid. Doors should be properly insulated.

Width of the corridor should be 15-20 feet.

Proper ventilators should be provided.

Proper artificaial light should be provided. Vertical tube lights should be provided.

Spawn lab (J)

It should be away from compost yard.

Built-up area is 80'x50'x15'(h).

The area should be divided into the following work area. They are space for office (A), Culture bank room (B), Store room (C), Toilet (D), Cold storage room (E), Laboratory (F), Washing room (G), Filling room (H), Boiling and Autoclaving Room (I), Inoculation room (J), Ante Room – Waiting room (K), Incubation room (L1-L2).

Post harvest handling room

Long term preservation methods are canning, pickling, freeze drying etc.,. About 500sq. feet area is required for these operations.

Chapter 15

SPAWN PRODUCTION

Seed of mushroom is called spawn. It is a mycelium growing on a substratum. Each piece of mycelium is developed in to different stages of mushroom. Spawn plays an important role in mushroom industry because success or failure of mushroom cultivation depends on the availability of pure culture of spawn.

Types of spawn

Natural spawn / virgin spawn

Mill track spawn

Brick spawn

Flake spawn

Manure spawn

Tobacco spawn

Grain spawn

Powder spawn

Granular spawn

Good spawn production needs the following materials and methods.

1. Layout of good spawn laboratory
2. Preparation and maintenance of pure culture
3. Preparation of stock or mother culture from pure culture
4. Preparation of spawn
5. Storage and transportation of spawn
6. Qualities of good spawn
7. Constrains and future strategy

Layout of good spawn laboratory

Size of the spawn laboratory is 80'x50'x15'(h) to handle 1000 spawn bottles.

Inoculation room

Incubation room

Media preparation room

Store room

Autoclaving room

Washing room

Filling room

Cold staorage

Culture bank room

Spawn lab is connected with water, proper drainage system and uninterrupted power supply.

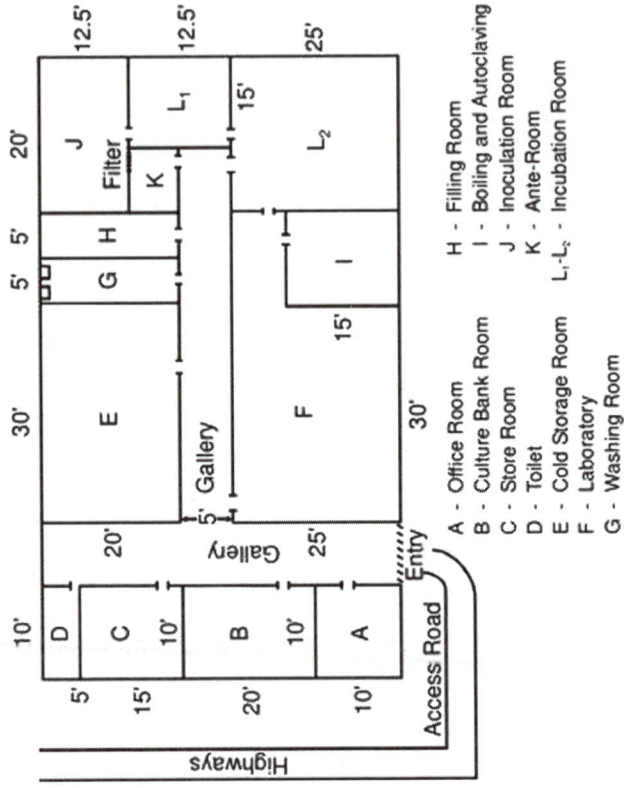

Fig.15.1 : Layout of Spawn production room.

Basic equipments needed

One or two autoclave or cooker

Incubator

Refrigerator

Laminar flow chamber

pH meter

Thermometer

Stove for boiling

Steam boiling

Test tubes and other glasswares

Working table

Shelves

Cooling system

UV lamp

Inoculation needle

Scalpels

Cotton

Polypropylene bag/ bottles

Rubber band

Labels

Aluminium foils

Chemicals – dextrose, agar agar, alcohol/ spirit, lactophenol cotton blue, formaldehyde, calcium carbonate, calcium sulphate, peptone, yeast extract, meat extract

Preparation of spawn

(i) **Starting Culture:** The mother culture can be obtained from any authorized agency or can be raised by any of the following three methods:-

(ii) **Spawn Media:** A number of materials, alone or in different combinations are popular as spawn substrates. The most common substrates are rice straw cuttings, sorghum, wheat & rye grains, cotton waste, used tea leaves etc. The protocols adopted for these substrates are mentioned below.

(iii) **Spawn containers:** Spawn containers vary from place to place depending upon the availability. Container selected should be heat resistance with cover, easily available in large numbers and should also be cheaper. Normally empty dextrose or saline bottles available in the hospital are used as spawn containers. Heat resistant plastic bags can also be used. They are light in weight and can be easily handled and transported.

(iv) **Mother spawns:** Mother Spawn can be used to inoculate either grain spawn or a second generation of mother spawn.

Preparation of grain spawn : For *Agaricus* spp. and *Volvariella* spp. only grain spawn is used. The main advantage of grain is that it is very nutritious for fungi and forms kernels easily. The kernels can easily be dispersed in the substrate. The main disadvantage is that it provides an optimal substrate for other organisms too. The chances of contamination are therefore high.

Grain spawn formula 1

Grain in small containers can be moistened to a higher degree than grain in 15 litre bags. For 2 litre containers, use the following recipe: 480 g rye, sorghum or wheat, 400 ml water, 2 g gypsum (45% moisture).

Grain spawn formula 2

Grain spawn substrate: grain 10 kg, $CaCO_3$ 147.5 g, Rice bran 1.25 g, Gypsum 0.1475g, Urea 0.5 g, Water 1.5 litres.

Preparation

About 100 kg grains are first boiled with about 150 liters of water for 20-30 minutes.

Spread grains on a sieve for 12-16 hours under shade.

Mix 2kg calcium carbonate and 2kg calcium sulphate with the surface dried cereal grains.

Fill glucose bottles or in polypropylene (PP) bags of 100 gauge thickness up to 2/3 portion. Close the mouth of the bottle using non-absorbent cotton plug. Similarly polypropylene bags are plugged with cotton after putting PVC ring.

Sterilize glucose bottles or PP bags containing spawn substrate at 126°C or 22 p.s.i for 2 hours followed by cooling under laminar flow bench under aseptic air.

Inoculate spawn substrate with mycelium culture followed by incubation at 32 + 2°C for about 2 weeks.

Storage of Pure Cultures and Spawn

The optimal temperature for growth of mushroom ranges from 30-35°C. Mycelium does not grow at all when the temperature is raised to 45°C or dropped to 15°C. At a temperature of 15±1°C most strains of mushroom survive for the longest period (Ahlawat, 2003). After complete colonization of spawn substrate with mushroom mycelia, it is ready to be used. However, if it is not to be used immediately, then it should be removed from incubator and stored at 15-20°C. At this temperature the growth of mycelium is arrested and mycelia are unharmed and remain viable for longer period (Ahlawat, 2003).

Purity of the spawn

Good spawn shows vigorous mycelial growth and contains no other organisms. If stored too long it will become less vigorous.

The final spawn

In order to inoculate the compost on the shelves (or the compost in the cultivation bags on the floor) larger quantities of spawn are used; generally referred to as final spawn. In order to prepare the final spawn, plastic bags can be used as spawn containers. The procedure for final spawn is similar to that of mother spawn. Only the sizes of the containers differ.

Purity of the spawn

1. Always prepare the spawn in a hygienic room. Avoid room with higher threatment (above 30°C). Dusty rooms and use of unsterilized materials should also to avoided.

2. The persons should maintain good hygiene.

3. Proper cooking and correct propagation (2%) of mixing of calcium carbonate to encourage good growth of spawn.

4. Sufficient inoculums should be given in the spawn bottles.

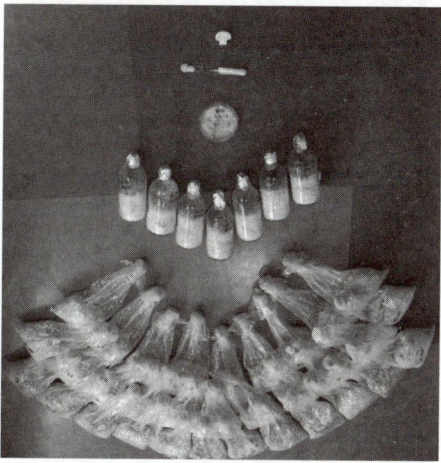

Fig.15.2 : The content of bags is ready to use the spawn.

5. Avoid loose plugging of spawn bottles.

6. Avoid frequent opening of the culture room.

7. Always keep the room free of dust and the flour should be clean with 1% dettol or other disinfectant.

8. Fumigation should be advised once in the month.

9. Aseptically transfer the mycelium to the bottles.

Methods of Spawning

The different methods followed for spawning are given below:

(i) **Spot spawning:** Lumps of spawn are planted in 5 cm. Deep holes made in the compost at a distance of 20-25 cm. The holes are later covered with compost.

(ii) **Surface spawning:** The spawn is evenly spread in the top layer of the compost and then mixed to a depth of 3-5 cm. The top portion is covered with a thin layer of compost.

(iii) **Layer spawning:** About 3-4 layers of spawn mixed with compost are prepared which is again covered with a thin layer of compost like in surface spawning.

The spawn is mixed through the whole mass of compost at the rate of 7.5 ml./kg compost or 500 to 750 g./100 kg compost (0.5 to 0.75%).

Chapter 16

COMPOST AND COMPOSTING PROCESS

Introduction

Compost is a mixture of organic and inorganic substances prepared from waste organic material by the way of microbial fermentations. Suitable substrate is needed for better mushroom cultivation. Compost is one of the most suitable substrate for mushroom cultivation.

Purpose of composting

It provides a medium.

It is favourable for mushroom spawn and able to predominate over other competitive mould.

Some of the nutrient already available is made more acceptable.

It increase protein content.

Heat generated during composting will damage unwanted microbes.

It will change physical nature of the substate.

Goals of composting

Suitable bulk density

Modification of plant materials

Biological removal of unwanted nutrients.

Building up of an appropriate biomass and a variety of microbial products.

Establishment of selectivity. It enhance mushroom growth.

Build up of compost moisture content.

Modification of compost structure.

Conversion of nitrogen into suitable organic matter.

Compost preparation

Traditionally, partially-decomposed horse-manure is used as a substrate for mushroom cultivation. But it is not enough for effective mushroom cultivation. Substrate plays a vital role in the yield of mushroom. Better substrate yields maximum yield. Hence compost is one of the most important substrate utilized for mushroom cultivation.

Materials used for composting

The materials used for compost preparation depend on the formulae used and method of composting. The commonly used materials are given below.

Base materials: Conventionally wheat straw either alone or mixed with horse-manure is the most widely used base material. Straws of the other cereals like rice or barely may also be used. The chief function of this is to provide cellulose, hemicellulose and lignin. These materials also provide proper physical structure to ensure the necessary aeration for the buildup of microbial population and the subsequent compost formaton. Rice and barley straws are quite soft and decompose quickly, leaving only a little fibre for imparting a proper physical structure to the compost.

Supplements:

Supplements are used for activating fermentation and can be categorised as:

(a) *Animal dung:* These include horse and chicken-manure. Nitrogen content may vary from 1 to almost 5%. In addition to nutrients, they contribute greatly to the final bulk density of the compost.

(b) Carbohydrate nutrients: molasses, wet brewers' grain and malt sprouts are the best source of carbohydrate.

(c) *Other nutrients:* Animal feeds, wheat or rice bran, dried brewer's grain, the seed meals of cotton, soya, castor and linseed are used to provide proteinand lipids. In these, both nitrogen and carbohydrate are available. Nitrogen content may vary from 3-12%. The oil and mineral content of some of these may be significant in mushroom nutrition.

(d) *Nitrogen fertilizers:* Nitrogen in chemical fertilizers (ammonium sulphate, calcium ammonium nitrate and urea) is rapidly released for the quick growth of microbial population.

(e) *Materials to correct mineral deficiencies:* Muriate of potash and calcium superphosphate.

Compost prepared from horse-dung mixed with straw are termed as 'natural', whereas they are called synthetic if the base material used is mainly straw without bulk animal-manure.

Formulations

There is no standard pattern in the compost formulations. However, 3 basic formulations for preparing compost are in use. The main objective of compost is being to achieve some of the balance between carbon and nitrogen. The nitrogen level of compost at stacking is adjusted to 1.5% of the dry matter and the carbon-nitrogen ratio at the same time is 25-30: 1. The compost should have 2.0-2.3% N at the completion of the process, which corresponds to 17:1, C-N ratio. There are so many variations in compost formulations. Some recommended formulae are:

1. Basic formula (IARI)	(in kg)
Horse dung	1,000
Wheat straw (chopped)	350
Urea	3
Gypsum (hydrated calcium sulphate)	30-40

Urea can be replaced with 100 to 110 kg of poultry manure.

2. Hayes and Randle (1969)	(in kg)
Horse dung	1,016
Chicken manure	101.6
Molasses	38.1
Cotton-seed meal	15.24
Gypsum	15
3. Synthetic compost Formulated at IARI, New Delhi	(in kg)
Wheat straw (chopped)/ Rice straw	1,000
Wheat bran	80
Urea	10
Ammonium sulphate or calcium ammoinum nitrate	10
Gypsum	40-50

Optional supplements. Molasses 40 kg or 20 kg molasses + 20 kg cotton seed or groundnut+seed meal; chicken manure 100-150 kg. Molasses should be diluted 20 times with water. Oilseed-meal cakes may be added during the first turning. Poultry-manure is added at the beginning of composting.

Method of composting

There are two methods for preparing mushroom compost, they are,

1. The long method: The 'long method' is considered primitive and unsuitable for commercial cultivation

2. Short method: The 'short method' is quick and has a definite advance over the earlier technology.

However, the 'long method' is still relevant for the growers in India who cannot afford the expensive technology required for the short method.

Composting yard : refer page No. 35.

Composting procedure

1. By long method

Wetting the straw: The first step in the composting process is to wet straw. The straw is spread thinly over the entire floor of the composting yard. It is then gradually wetted by sprinkling water, gently, till the straw takes no more water. The straw is then turned for even wetting. Again water is sprinkled till it can absorb no more. One ton of dry straw will require almost 5,000 litres of water to bring it into saturation.

Mixing and heaping (*Fig.16.1*): After the straw is wetted, the supplements excluding the gypsum are uniformly scattered over the straw and mixed. After mixing, the mixture is finally stacked in a heap. A heap one meter height, one meter wide and of indefinite length has been found to be suitable. The straw can be stacked manually or with a stack mould. The straw should be firmly but not compactly compressed into the mould. The dimensions of the heap

Fig.16.1 : Composting : Out door method

can be adjusted according to the size of straw and air temperature. The principle is that longer the straw, bigger the heap. If composting is done in the cooler months when the temperature ranges between 10° and 18°C, a small heap would be unable to retain heat and moisture and the composting would be unsatisfactory. During the hot weather generally and in particular in tropical and sub-tropical regions, the temperature difference between inside of the compost and the surrounding air is too small to produce chimney-effect necessary for compost ventilation. As a rule undesirable acid zones occur inside the compost. In such cases, relatively narrow heaps would be more suitable.

Turning schedule: It is important to ensure that the heap attains sufficiently high temperatures (70°-75°C) to bring about the correct composting; otherwise the compost will lack the necessary nutritive value. Care must also be taken to see that over composting does not take place. Open the heap and make it a number of times and for this purpose, the time schedule suggested is:

Day zero: Wet, mix the stack and the heap

4th day: First turning

8th day: (Second turning) the pile is opened and turned. Watering may be done to dry patches.

12th day: (Third turning) the pile is opened and required   quantity of gypsum is added and restacked.

16th day: (Fourth turning) Cotton seed meal is added.

20th day: (Final turning) Care should be taken and watering is done if necessary.

28th day: Filling of the trays.

Nitrogeneous supplements and carbohydrates are mixed on day zero. Gypsum is usually mixed at the third and forth turning in quantities. During the final turning, 40 ml Malathion diluted in 20 litres of water is sprinkled. Any other available insecticide, like DDT, BHC or Lindane can also be used. The guiding principle is that the heap should be opened when the temperature within rises no further.

Fig.16.2 : Composting : Indoor method.

2. By short method

The method which was developed by Sinden and Hauser (1950) constitutes a general advance in controlled composting. The short method consists of two phases- Phase I and phase II.

Formulae

1 Formula-1 (College of Agriculture, Solan)

Wheat straw	1000 Kg
Chicken manure	400 Kg
Brewers grain	72 Kg
Urea	14.5 Kg
Gypsum	30 Kg

2. Formula-2 (College of Agriculture, Solan)

Wheat straw	500 Kg
Horse manure	1000 Kg
Chicken manure	300 Kg
Urea	7 Kg
Gypsum	30 Kg
Brewers grain	60 Kg

3. Formula-3 (Indian institute of Horticultural Research, Hassaraghtta)

Wheat straw	300 Kg
CAN or Ammonium sulphate	9 Kg
Super phosphate	9 Kg
Wheat bran	15 Kg
Gypsum	30 Kg

4. Formula-4 (Indian Agricultural research Institute, New Delhi)

Wheat straw	350 Kg
Horse manure	1000 Kg
Chicken manure	100-150 Kg
Urea	3 Kg
Gypsum	30-40 Kg.

5. Formula-5 (Mushroom Research Laboratory, Solan)

Wheat straw	1000 kg
Urea	14.5 kg
Chicken manure	400 kg
Brewers grain	72 kg
Gypsum	30 kg

Commercial Mushroom Substrate Preparation (Composting) Process

Phase I: Making mushroom compost (taking place in the open air)

Raw materials preparation:

Straw-bedded horse manure and hay or wheat straw are the common bulk ingredients. The only inorganic supplement left is gypsum which is of utmost importance (gypsum may be added early in the composting process, at 70–100 lbs per ton of dry ingredients).

A suggested formula for trial use is as follows (the following mentioned formulation only for reference):

The whole composting begins by mixing and moistening base materials.

First, soak the wheat straw in water and crush it through crusher machine. The moisture should be controlled moderate, not too wet or too dry. There must be adequate moisture, oxygen, nitrogen, and carbohydrates present throughout the compost pile. Too low temperature or moisture will affect the process, even stop the composting.

Second, blend the chicken(horse) manure and gypsum together and dropped into the crushed straw.

Finally, these mixed materials are delivered to composting piles for composting.

Composting

Building the raw ingredients into long rectangular piles and Periodically turning the piles (about 3-4 turning on each 2-3 days to allow the manure to rot down and concentrate the nutrients required for mushroom growing). A compost turner plays an important role in commercial composting. It is to mix the ingredients thoroughly, accelerate the chemical reaction during composting and water the ingredients (add water to run-off). A tractor-loader to move the ingredients to the turner is also needed.

Phase I composting lasts from 7-14 days depending on the condition of the material at the start and its characteristics at each turn. It is considered complete when: (a) the mixture is dark brown and sweet smelling, (b) straws become soft & pliable and lumps break apart easily, (c) raw ingredients are capable of holding water, and (d) the moisture content of the compost is from 68-74%. When the moisture, temperature, color, and odor described have been reached, Phase I composting is completed.

Phase II: Finishing the compost (occurring under controlled conditions)

After 15-20 day's composting, the Phase II process to begin, staring with a pasteurization to kill bacteria, weed seeds and remove the ammonia.

Compost is delivered into a special room to have a pasteurization period of 8h at 56-60°C and continues with a conditioning period at 45°C for up to 7 days until volatile NH3 has been cleared from the process air. The pasteurization process is a controlled, temperature-dependent process that lasts for about one week. After Phase II, the substrate is ready for growth of mushroom mycelium.

Chapter 17

CASING

Mushroom bed must be covered with a suitable material to induce the change from vegetative growth of mycelium to fruiting body are called casing. Bed covering layer is called casing layer. The process of applying casing layer is called casing.

This practice was developed by *Agaricus* growers who found that mushroom formation was stimulated by covering their compost with such a layer. A casing layer encourages fruiting and enhances yield potential in many, but not all, cultivated mushrooms.

Functions of casing layer

The basic functions of the casing layer are:

It protect the colonized substrate from drying out. It provide a humid microclimate for primordia formation and development. It provide a water reservoir for the maturing mushrooms. It support the growth of fructification enhancing microorganisms.

Properties of casing layer

The casing layer must maintain mycelial growth, stimulate fruiting and support continual flushes of mushrooms.

Water Retention: The casing must have the capacity to both absorb and release substantial quantities of water.

Structure: The structure of the casing surface must be porous and open, and remain so despite repeated waterings.

Microflora: High levels of bacteria such as *Pseudomonas putida* result in increased primordia formation, earlier cropping and higher yields. Dense casing layers and deep casing layers generally yield more mushrooms because they slow diffusion.

Nutritive Value: Have low nutritional value compared to the substrate. A nutritive casing supports a broader range of competitor molds.

pH: The pH of the casing must be within pH values between 7.0-7.5.

Hygienic Quality: The casing must be free of pests, pathogens and extraneous debris.

Quality of casing materials

- Soft texture
- Light weight
- High water holding capacity
- High porosity
- Deficient in available form of C and N
- Neutral pH (7.0 – 7.5)
- Low conductivity (400-600 ì moh)

Casing Layer

Spawn

Compost

Fig.17.1 : Casing.

Casing materials

Earlier sub-soil material or organic matter rich soils were used as casing in button mushroom cultivation. Presently peat is the most desirable casing material used world wide with excellent mushroom yields and superior fruit body quality. However, peat is not available in India. The other alternative recommended materials are,

Sand: Characterized by large individual particles with large air spaces in between, sandy soils are well aerated. Their structure is considered "open". Sandy soils are heavy, hold little water and release it quickly.

Clay: Having minute individual particles bound together in aggregations, clay soils have few air pockets and are structurally "closed". Water is more easily bound by clay soils.

Loam: Loam is a loose soil composed of varying proportions of sand and clay, and is characterized by a high humus content.

- Well decomposed Farm Yard Manure (FYM) preferably two years old
- Well decomposed Spent Mushroom Compost (SMC) (two years old anaerobically decomposed)
- Composted coir pith (coir industry waste) (well decomposed & water leached)
- 1:1, 2:1 and 1:2, v/v of well decomposed FYM and SMC
- 1:1, v/v of decomposed FYM or SMC with composted coir pith
- Decomposed powdered bark of some forest trees
- Paper industry waste
- Burnt rice husk is also in use along with decomposed FYM (2:1, v/v) in seasonal cultivation of button mushroom in Hayrana and Punjab with reasonable success.

Casing Formulas and Preparation

Farm yard manure with loam soil of 1:1 ratio is considered as a best casing medium. Suman 2004 reported that farm yard manure+coconut coir pit in the ratio of 4:1 as a best casing medium for Agaricus.

Highest button mushroom yield was obtained on casing the mycelial colonized substrate with the casing material prepared with FYM + spent compost + soil + sand in the ratio 1:1:1:1, followed by FYM+burnt rice husk (1:10), vermi compost + loam soil (1:1), spent compost and farm yard manure at Coimbatore Centre. Compost formulation with sugarcane baggase + wheat straw (2:1) resulted in reduced cost of cultivation of Agaricus Bisporus.

Disinfection of casing material is by the use of formaldehyde or steam. It removes harmful mould, eelworm and mits. Spray one liter of 40% formaldehyde to one cubic meter of casing medium. Steaming for 5-6 hours at 60-65°C or 4 hours at 70-75°C.

Casing and case run

Casing is a 3-4 cm thick layer of soil applied on top of spawn run compost and is a pre-requisite for fructification in A. bisporus.

- Make a heap of casing material
- Wet it up to 50-60% water holding capacity
- Fill in trays and shift them to pasteurization chamber
- Steam pasteurization at 60-65°C for 6-8 hours
- Auto-Cooling
- Unfold the fully spawn run bag and make the top surface even by gentle pressing with hands
- Light spray of water on spawn run compost
- Application of 4-5 cm thick layer of casing uniformly.
- Water spray in installments immediately after casing application

Within three days of application, the mycelium should be growing into the casing layer. Once mycelial growth is firmly established, the casing is gradually watered up to its optimum moisture holding capacity. This is accomplished by a series of light waterings with a misting nozzle over a two to four day period. IT IS EXTREMELY IMPORTANT THAT THE WATERINGS DO NOT DAMAGE THE SURFACE STRUCTURE OF THE CASING

Case run and pinhead formation

Case run is done at a temperature of $24 \pm 1°C$, RH-95% and $CO_2 > 7500$ ppm (strain dependent) for about one week. There is no requirement for fresh air introduction during case run. It is considered complete when mycelia come in the valleys of casing layer. After case run, the environmental conditions are changed by bringing down the temperature to 15-17°C (air), RH to 85% and

CO_2 to 800-1000 ppm (strain dependent) by opening of the fresh air ventillation and exhausting CO_2. This change in environmental parameters induces pinhead formation in 3-4 days (strain dependent) time. The pinheads develop into solid button sized mushrooms in another 3-4 days. At this stage, the air inside the cropping room is changed 4-6 times in an hour to maintain appropriate CO_2 conc. as CO_2 production is at its peak during first flush (actually peaks at case run).

Chapter 18

CULTIVATION TECHNOLOGY OF THE WHITE BUTTON MUSHROOM (*AGARICUS BISPORUS*)

Introduction

Agaricus bisporus is the scientific name for button mushroom. It is otherwise called as Pizza mushroom. *Agaricus* is generally called as gilled mushroom. *Agaricus* is a saprophytic mushrooms with a chocolate brown spore print. *Agaricus bisporus* produces only two basidiospores on each basidium. *Agaricus bisporus* is the most commonly grown mushroom in Asian countries. *Agaricus bisporus* has increased in popularity in North America with the introduction of two brown strains, Portabella and Crimini.

Taxonamic position (*Fig.18.1*)

Kingdom : Fungi

Division : Bacidiomycota

Class : Agaricomycetes

Order : Agaricales

Family : Agaricaceaceae

Genus : Agaricus

Species : Agaricus bisporus

Structure of Agaricus

1. The vegetative mycelium is composed of many inter-woven sepatate hyphae.

2. The reproductive phase is initiated by the formation of small knob like swellings at different points of interwoven mycelial strands.

3. These swellings increase in size and break through the surface of the substratum as small balls constituting the button stage.

4. A matured basidiocarp (fruit body) is white in colour and consists of thick short stipe with an annulus.

5. The stipe supports the pileus which appears as a hat like expansion.

6. On the underside of the pileus, a number of radiating gills or lamella are present which are pink when young but purple-brown when mature.

External Structure and Internal Structure: *Refer Chapter 8.*

Cultivation method

Agaricus is a secondary decomposer. Hence it grows effectively in compost. White button mushroom is a temperate mushroom which requires low temperature.

Fig.18.1 : Spawn culture.

Compost preparation

The white-button mushroom is grown on a selected substrate which provides adequate levels of nutrients to support the crop. Traditionally, partially-decomposed horse-manure has been the principal medium for providing the required nutrients in artificial cultivation of the mushroom and it is only in recent times that other materials have also been used successfully.

Refer : Chapter No. 16.

Substrate

Compost is a major substrate taken for button mushroom cultivation. Complete compost is filled in trays for mushroom cultivation. Compost is filled in a trays.

Spawning

The process of mixing spawn with compost is called spawning. Commercial mushroom growers obtain spawn from any spawn companies. Farmers have a choice of growing different strains, ranging from smooth white, off-white, cream, to brown capped mushrooms. These strains vary in flavour, texture, and growth requirements.

Casing

The important transition stage from the vegetative to the reproductive stage of *A. bisporus* takes place in the casing layer. Mushrooms form only after the compost is enclosed with a layer of casing material. The casing layer provides moisture important for high yields and anchorage for the developing mushrooms. Casing materials do not provide any nutrients to the mushroom mycelium. Environmental conditions after casing are the same as during spawn

*Fig.18.2 : Mushroom cultivation. 1. Spawning process,
2. Panning, and 3. Fruiting*

growth. The compost temperature is kept around 24oC for up to 5 days after casing to allow for the spawn to grow through the casing layer. Before the mushroom "pins" start to develop, later is applied intermittently to raise the moisture level of the casing layer to field capacity. Most *Agaricus* is grown in a place with high relative humidity and not much light.

Pinning

Primordia or "pins" are knots of mycelium that eventually develop into mushrooms. Once the mycelium has reached the surface of the casing layer, the mushroom is induced to pin by sinking both the air temperature (to 16-18oC) and the CO_2 concentration (to 0.08%). Fruiting occurs in well-defined flushes or breaks with the first harvestable mushrooms appearing 18 to 22 days after casing.

Cropping

The mushroom crop grows in repeating 3- to 5-day cycles called "flushes" or "breaks". These flushes are followed by a few days when no mushrooms are available to harvest. The individual flushes tend to produce progressively fewer mushrooms. Most mushroom farmers crop their mushrooms for 30 - 40 days. During cropping, the casing layer is watered 2 to 3 times per week and air temperatures are maintained between   15-18oC. This temperature range favours mushroom growth and lengthens the life cycles of both disease pathogens and pests. An entire production cycle, from composting to final harvest, can take up to 15 weeks.

Fruiting

Under favourable environmental conditions viz. temperature (initially 23 ± 20°C for about a week and then 16 ± 20°C), moisture (2-3 light sprays per day for moistening the casing layer), humidity (above 85%), proper ventilation and CO_2 concentration (0.08-0.15 %) the fruit body initiate as pin heads start growing and gradually develop into button stage.

Harvesting and Yield

Harvesting is done at button stage and caps measuring 2.5 to 4 cm across and closed are ideal for the purpose. The first crop appears about three weeks after casing. Mushrooms need to be harvested by light twisting without disturbing the casing soil. Once the harvesting is complete, the gaps in the beds should be filled with fresh sterilized casing material and then watered. About 10-14 kg fresh mushrooms per 100 kg fresh compost can be obtained in two months crop. Short method used for preparation of compost under natural conditions gives more yields (15-20 kg. per 100 kg. compost).

Chapter 19

CULTIVATION OF OYSTER MUSHROOM

Introduction

The oyster mushroom is a common edible mushroom. Scientifically this mushroom is called *Lentinus sajorcaju* (Synonymous. *Pleurotus sajorcaju*). Oyster mushrooms can also be used industrially for mycoremediation purposes. Many believe that the name is fitting due to the flavour resemblance to oysters. In Chinese, they are called "flat mushroom". Traditionally it is cultivated in Ascian countries. Now it is cultivated throughout the world.

*Fig.19.1 : **Gill structure of oyster mushroom.***

Taxonamic position

Kingdom : Fungi

Division : Bacidiomycota

Class : Agaricomycetes

Order : Polyporales

Family : Polyporaceae

Genus : Lentinus (Syn. Pleurotus)

Species : *Lentinus sajorcaju*

Morphology

Cap: 5-25 cm broad, fan or oyster-shaped; Natural specimens range from white to gray or tan to dark-brown; margin inrolled when young, smooth and often somewhat lobed or wavy. Flesh white, firm, varies in thickness due to stipe arrangement.

Gills: Gills are white to cream, descend stalk if present. If so, stipe off-center with lateral attachment to wood.

Spores: The spores form a white to lilac-gray print on dark media.

Stipe: Often absent. When present it is short and thick.

Taste: Mild

Odor: Often has a mild scent of anise.

Mushroom cultivation

There are several Pleurotus spp.,(viz., *P. ostreatus, P. flabellatus, P. sajorcaju, P.florida, P.cornucopiae, P. citrinopileatus, P. cystidiosus, P. fossulatus, P. opuntiae, P. tuberangium, P. opuntiae, P. membranaceus, P. oryngii* etc). Among these P. sajor caju species is most predominantly cultivated in thatched, polythene, brick or stone houses. These mushrooms are well-known for their high bioeffeciency percent, delicacy and flavour. Oyster mushroom forms protein rich fruiting bodies from the cheapest substrate.

Preparation of substrate

Oyster mushroom can be grown on various substrates. Many agricultural waste and by-products like cereal straw, hulled maize cob, tapioca starch waste, unused waste paper, rubber wood waste, waste cotton, panana pseudostem, wood shavings, saw dust, paddy straw, maize stalks/cobs, vegetable plant residues etc. Since paddy straw is easily available and cheap, it is widely used. Paddy straw should be fresh and well dried.

Soaking: Chop paddy straw into 3-5 cm pieces and soak in fresh water for 8-16 hours. If maize stalks/cobs are used, soaking period should be 24-48 hours. Drain off excess water from straw by spreading on raised wire mesh frame.

Heat treatment: Heat treatment of substrate results in minimizing contamination problem and gives higher and almost constant yields. It can be done in two ways i.e. by pasteurization and sterilization by chemicals.

(i) **Pasteurization:** Boil water in a wide mouth container such as tub or drum. Fill the wet substrate in gunny bag or basket and close the opening. Dip the filled bag in hot water of 80-85o C for about 10-15 minutes. To avoid floating, press it with some heavy material or with the help of a wooden piece. After pasteurization, excess hot water should be drained off from container so that it can be reused for other sets. Care should be taken to maintain hot water temperature at 80-85oC for all sets to achieve pasteurization.

(ii) **Chemical sterilisation technique:** Take 90 litres of water in a drum of 200 litre capacity. Slowly steep 10 kg of chopped paddy straw in the water. Mix 125ml of formaldehyde (37-40 percent) and 7 g of Bavistin dissolved in 10 litres of water in another container and pour the solution slowly into the drum. Straw should be pressed and drum should be covered with a polythene sheet. Take out the straw after 12 hrs. Spread the pasteurized or chemically sterilised straw on neat and clean cement flooring or on raised wire mesh frame, inside the chamber where bag filling and spawning are to be done.

Preparation of mushroom beds: In the cultivation of oyster mushroom out of different methods of cultivation used, cylindrical bed method is found to give higher yields with lesser cost. Pasteurized substrate is used for filling and spawning. Substrate moisture content should be about 70% and 25-28°C. Polythene bags (35 x 50 cm, 150 gauge) or polypropylene bags (35 x 50 cm, 80 gauge) may be used for its cultivation. One 500 ml bottle spawn (200-250 g)

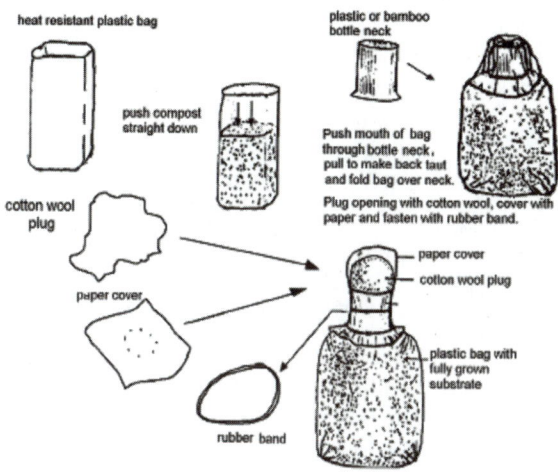

Fig.19.2 : Steps in oyster mushroom cultivation.

can be used for filling 3 bags of straw. Spawning can be done in layer spawning. Fill the substrate in bag, press it to a depth of 8-10 cm and spread a handful of spawn above it. Similarly, 2nd and 3rd layers of substrate and simultaneously after spawning, the bags should be closed. After gentle pressing, close the bags for spawn running (development). The bag is generally filled with 5 layers of straw bits and four layers of spawn. Tie the closed end of the bag with jute and revert it to get the inner side as outer side. This provides a flat bottom to the bag.

Spawn running

Spawned bags should be stacked in racks in neat and clean place, in closed position. Temperature at 25°C and humidity at 70-85% should be maintained by spraying water twice a day on walls and floor. The mushroom mycelium in the spawn grows on the paddy straw pits inside the polythene bag. The growth of the mushroom fungus inside the bed may be watched every day. No light is necessary during spawn run. The fungus will exhibit white growth of the mycelium on the straw bits and it grows slowly on the surface of the straw and covers the entire bed. It takes 20-22 days when bags will be fully covered with white mycelium. The time may vary depending on the type of spawn, substrate, and temperature of the room, moisture content of the bed and sterilization methods. After noticing the full growth in the bag, take out the bag and cut open the polythene bag along the length with the help of a sterile knife or blade and remove the polythene covering. This is made on 16th to 21st day in white oyster mushroom and gray oyster mushroom respectively. After removing the polythene bags a compost cylindrical bed tightly intertwined by the mycelia.

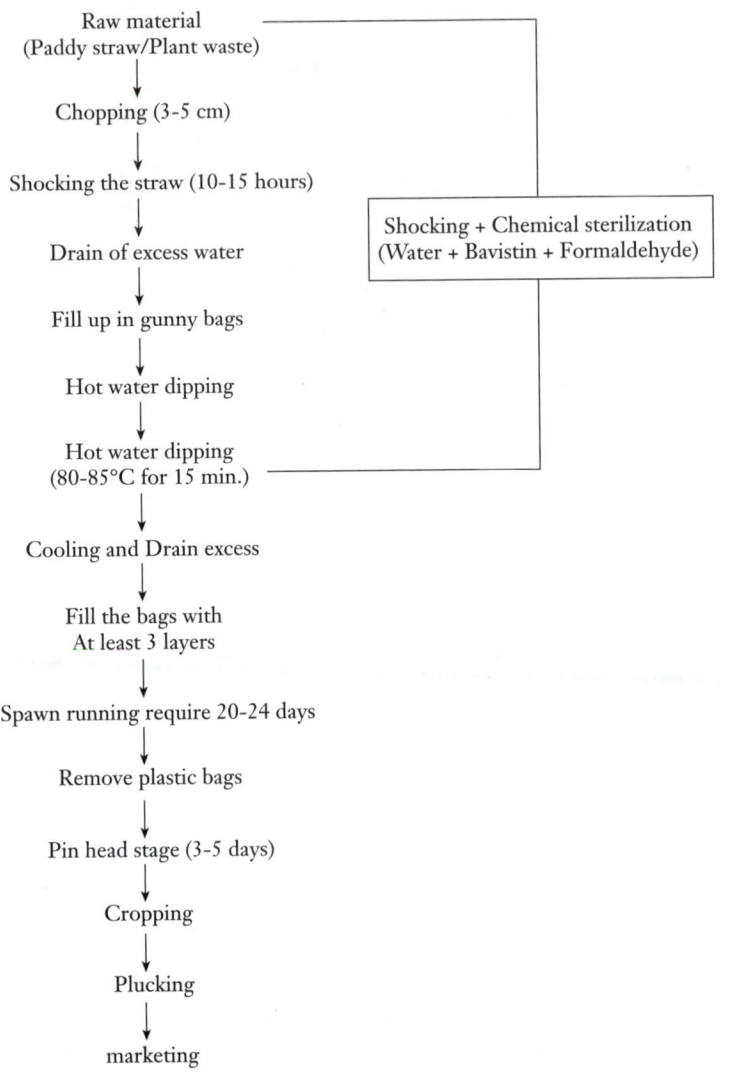

Fig.19.3 : Flow chart for mushroom cultivation.

Cropping and Harvesting

After 20-22 days, when bags are fully impregnated with white mycelium, transfer the bags into cropping room and remove polythene/ polypropylene covers. The open blocks should be kept in racks about 20cm apart. Rack should be 60 cm wide with gap of 50-60 cm between two shelves. Mushrooms grow in a temperature range of 20-33oC. Relative humidity is maintained by spraying water twice a day on the walls and floor of the room. Spraying of blocks should be avoided for the first 2-3 days. A light mist spray of water is given on blocks

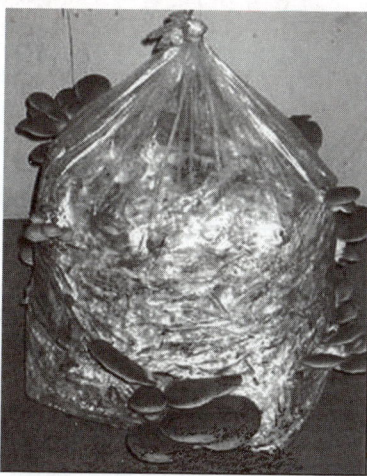

Fig.19.4

as soon as the small pin heads appear. Once pinheads are 2-3 cm big a little heavier watering is to be done on blocks and further watering of blocks is to be stopped to allow them to grow.

Mushrooms should be plucked before they shed spores to maintain quality. After 1st flush of harvest, 0.5 to 1cm outer layer of the block should be scrapped. This helps to initiate 2nd flush which appears after about 10 days.

After harvest, the lower portion of the stalk must be cleaned with dry cloth. They should be packed in perforated (5-6 small holes) polythene bags to keep them fresh. It loses freshness after about 6 hours, which can be enhanced by keeping them in refrigerator. Oyster mushroom can be sun dried for 2 days and dried product marketed in polythene bags. Dried mushrooms should be soaked in water for 10 minutes before use.

Yield of mushroom: From each bed 650-750g of mushroom can be harvested. Mushroom of 400-500g from the first harvest, 150g from the second harvest and 100g from the third harvest. Better quality mushrooms are obtained from first harvests compared to second or third harvests.

Chapter 20

CULTIVATION OF PADDY STRAW MUSHROOM

Introduction

Volvariella volvacea is called as Paddy straw mushroom. It is also known as the Chinese mushroom. It belongs to the family Pluteaceae of Basidiomycetes. It is an edible mushroom. It is cultivated in tropics and subtropics regions. It is first cultivated in China in 1822 and India in 1940. It is also known as "warm mushroom" as it grows at relatively high temperature. It is a fast growing mushroom under favourable growing conditions. The total crop cycle is completed within 4-5 weeks. It can be grown on uncomposted substrates such as paddy straw, cotton waste or other cellulosic organic waste. It has been considered as one of the easiest mushrooms to cultivate.

Taxonomic position

Kingdom : Fungi

Division : Bascidiomycota

Class : Agaricomycetes

Order : Agaricales

Family : Plutaceae

Genus : Chamaeta; Pluteus; Volvariella

Biological Characteristics

The fruiting body of this mushroom is divided into six different developmental stages. They are Pinhead, tiny button, button, egg, elongation and mature stage.

1. **Pinhead stage:** It is of the size of a pinhead. Colour of the veil is perfectly white (*Fig.20.1*). The whole structure is a knot of hyphal cells.

2. **Tiny button:** It is formed from interwoven hyphae. Only the top of the veil is brown in colour and the rest is white (*Fig.20.1*). It is round in shape. The lamellae are seen as a narrow band on the lower surface of pileus.

3. **Button stage:** This stage is sold in the market at a premium price. In this stage, the whole structure is wrapped by a coat, which is called as the universal veil (*Fig.20.1*). Inside the veil, closed pileus exists.

4. **Egg stage:** Pileus is pushed out of the veil and the veil remains as volva (*Fig.20.2*). The stipe is again not visible at this stage. The lamellae of this stage do not bear basidiospores. The size of the pileus remains very small upto this stage.

5. **Elongation stage:** The pileus remains close and the size is smaller than mature stage, while the stipe attains the maximum length (*Fig.20.2*). The stipe is marked with water proof drawing ink.

Fig.20.1 : Developing mushrooms on plantain leaf bundies
(13 days after spawning).

6. **Mature stage:** The pileus is connected in the centre with stipe and is of usually 6 to 12 cm in diameter. The fully grown pileus is circular in shape with an entire margin and smooth surface. The surface is dark grey at centre and light grey near the margin. The lamellae vary in size from full size to one quarter size of pileus. The colour of basidiospores again vary and it may be of light yellow, pink or dark. Another important part of mature fruiting body is the stripe, which connects the volva and the pileus.

Lifecycle

Refer Page No. 13.

Nutritive Value

This mushroom is distinguished from others by its unique flavours and textual characters. It contains around 90% water, 30-43% crude protein, 1-6% fat, 12-48% carbohydrates, 4-10% crude fibre and 5.13% ash. This mushroom is rich in potassium, sodium and phosphorus. Potassium constitutes the major fraction of major elements followed by sodium and calcium. It also contains minor elements like Cu, Zn and Fe. Vitamins like thiamin and riboflavin are also available in higher concentrations. Various proportions of lysine, glutamic acid, aspartic acid, tryptophan, methionine and phenylalanine are available.

Mushroom production

Paddy straw mushroom prefers high cellulose, low lignin containing substrate. Waste like paddy straw, water hyacinth, oil palm bunch, oil palm pericarp waste, banana leaves, saw dust, cotton waste, sugarcane bagasse etc. contains high cellulosic contents. The cultivation of Volvariella is less sophisticated, less extensive and can be rewarding in tropical & subtropical climates.

Various methods are available to cultivate paddy straw mushroom.

Conventional Method

The different steps involved in this method are as follows:

- This mushroom grows best on paddy straw. The straw is tied into bundles of 1.2m Long x 25cm dia.

- Bundles are immersed in clean water for 12-18 hours in a cemented water tank.
- Excess water is drained by placing bundles on raised bamboo platform.
- Straw bed is prepared by placing 4 bundles side by side and another four bundles similarly but from the opposite side, forming one layer of eight bundles. The open ends of bundles from opposite sides should overlap in the middle.
- Second, third and fourth layer are formed by by intermittent spawning between first and
- second, second and third and third and fourth layers.
- Entire surface of the layers of the beds are spawned at a space of 5cm apart leaving margin of 12-15cm from edges.
- Red gram powder is sprinkled over the spawned surface. (Use 500 gm spawn and 150 gm of red gram powder for a bed of 30-40 kg of dried paddy straw).
- Straw bed is pressed firmly from the top and covered with clean plastic sheet for maintaining required humidity (80-85%) and temperature (30-35°C).
- Plastic sheet is removed after 7-8 days of spawning and maintaining temperature of 28-32°C and relative humidity about 80%.
- Mushroom will start appearing after 4-5 days of sheet removal and will continue for next 20 days.
- Substrate after harvest can be used as manure.
- Each bed of 30kg dry straw can produce 5-6 kg of fresh mushrooms.

INDIAN VILLAGE METHOD

Materials required

Polythene bags, straw, spawn, drum for water boiling,

Method

Collect good quality straw.

Cut the straw in to pieces (Chopping).

Sterilize straw by immersing in boiling water for 2-4 hour (keep straw in gunny bag and immerse in water – easier method).

Drain off excess water.

Fill the polythene bag with the layer of straw and a layer of spawn (Layer spawning). Fill upto 2/3 potion of the bag.

Tie bag using thread.

Spawn run the filled bags by placing the bag in a racks with sufficient space and environmental conditions (RH-80-85% and Temperature 25-30°C).

Mushroom grow as a mycelia mass (After 10 days remove polythene bag)

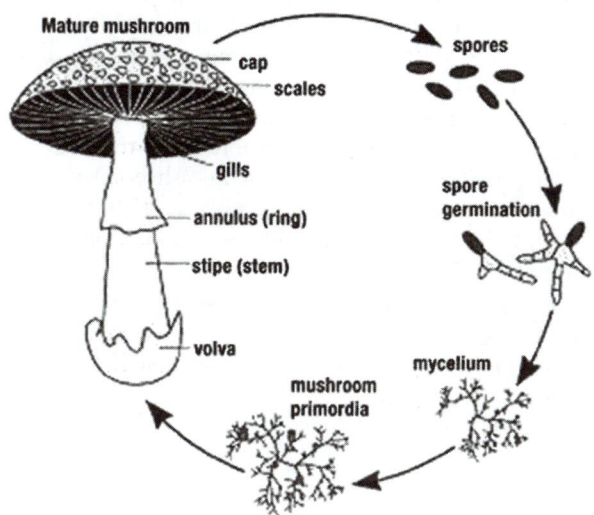

Fig.20.2 : Life Cycle.

Pinhead appears after 12 days.

After 15 days firs flesh will appear.

Four to five flues may be harvested

Harvesting and Processing

Harvesting : The straw mushroom is harvested before the volva breaks or just after repture. These stages are called as the button and egg stages. This mushroom grows at high temperature and moisture, therefore, its growth is very fast.

The mature fruiting bodies should be carefully separated from the beds/ substrate by lifting and shaking slightly left or right and then twisting them off. The mushrooms should not be cut off by knives or scissors from the base of the stalk, because the stalk left behind on the bed/substrate will rot and be attacked by pests and contaminated by moulds, which in turn will destroy the mushroom bed.

Processing : Straw mushroom is more perishable than other edible mushrooms and can not be stored at 4°C as it undergoes autolysis at this temperature (Ahlawat et al. 2006). This mushroom can be stored at a temperature of 10 to 150C for 3 days and little more at 20°C or under controlled atmosphere storage. The loss of moisture in 4 days stored mushroom could be as high as 40-50% in unpacked mushroom, while it can be reduced to 10%on packaging in perforated polythene begs. Straw mushroom can be processed by canning, pickling and drying.

Air Drying : Sun drying is very common in straw mushroom. The mushrooms are cut longitudinally before drying. Drying by hot air is better than sun drying because mushroom retains better flavour and colour. Drying takes place in 24 hours at 30°C. However, mushroom can also be dried with temperature beginning at 400C than increasing gradually until it reaches at 450C for eight hours. The optimum drying temperature, time and critical moisture content for drying of the paddy straw mushroom has been recorded tobe 60°C and 7 hrs. Fresh mushrooms are reduced to about one-tenth of their original weight after dehydration. Dried mushrooms should be placed in air tight containers, to prevent moisture absorption. Dried mushrooms can be powdered and then used for making soup, ketchup or curry after reconstitution in water.

Freeze Drying: Freshly picked mushrooms are to be frozen at – 20°C and then freeze dried. The finished produce on rehydration used to be better than air-dried product. On reconstitution it becomes almost indistinguishable in appearance from the fresh ones.

Chapter 21

CULTIVATION OF SHIITAKE MUSHROOM

Introduction

Scientifically this mushroom is called *Lentula edodes*. It is native to East Asia. It is considered as medicinal mushroom. It is also a saprophytic fungus. The cultivation of shiitake is rapidly gaining popularity. In nature, this mushroom is growing on wood legs. On commercial purpose, it is cultivated on bags using sawdust. In this bag method, Shitake mushroom grow fast and gives higher yield when compared to wood logs.

Taxonamic position

Kingdom : Fungi

Division : Bascidiomycota

Class : Agaricomycetes

Order : Agaricales

Family : Marasmiaceae

Genus : Lentula

Species : *L. edodes*

Importance of Shitake mushroom

* Protect DNA from oxidative damage.
* Eritadenine helps to reduce cholesterol and helps in fight obesity.
* Linoleic acid helps weight loss and building muscle.
* Support immune functions.
* Destroy cancer cells.
* Support cardiovascular health.
* Boosts energy.
* Improves brain function.
* Have antimicrobial properties.
* Provide Vitamin D
* Reduce the risk of diabetes
* Promote skin health.

Nutrients

Carbohydrate - 75.37g

Fat - 0.99g

Protein - 9.58g

Selinium - 46mg

Cultivation of mushroom

It is cultivated commercially using sawdust bags.

The main advantages of growing shiitake on bags are:

1. Many types of organic waste can be used.

2. Total cropping period is 6 months compared to 4 to 6 years with cultivation on wood logs.

3. The substrate has been compressed and only little spawn has been used.

Substrate for mushroom cultivation

The most commonly used substrate formulations are:

• Sawdust, 3 to 4% rice bran, 1% corn meal or wheat bran, 1% $CaCO_3$

• Sawdust, 10 to 25% corn waste, 1 to 2% $CaCO_3$

1. Fresh sawdust from the trees of the oak, sweat chesnut.

2. Sawdust from other trees can also be used, but if the sawdust contains resins it has to ferment for a number of (up to 6 months). When the sawdust is moist enough it has to be mixed with the supplements and the chalk.

3. Mix the chalk first with the rice bran, as it is easier to get an even distribution. The moisture content at the time of preparation is usually between 55-65% of the substrate and increases during incubation.

Filling: Fill polythene and polypropelene bag with 1.2kg of saw dust substrate. Streaming under low pressure is appropriate method

Growth stages: Five different room temperature stages of mycelial growth of siitage can be distinguished for all strains. The first phase is the normal spawn run as it occurs in all fungi. When the substrate has turned white, it is not ready to fruit. It has to mature first.

Spawning and spawn run: After steaming, the bags are cool down and spawn them the next day. 10 g of sawdust spawn is sufficient to spawn one bag of 1.2 kg saw dust. Spawn run will take one to four months for the mycelium to colonise the substrate and mature. For fruiting some light should be provided for at least the end of the spawn run. All strains show optimal mycelial growth at 25°C. The temperature inside the bags is usually a few or even ten degrees higher than the ambient room temperature.

The following are the five stages:

(a) **Mycelial growth:** The spawn will give rise to white hyphae, which produce enzymes to degrade complex substances like cellulose, lignin and hemi cellulose into smaller fragments. The fragments will be consumed at later stages of mycelial growth.

(b) **Coat formation:** A thick, white mycelial sheet will develop on the surface of the substrate. This will occur in two to four weeks after inoculation. If the CO_2 level is high, the sheet will be thicker.

(c) **Mycelial clumping:** Commonly formed on the surface by most strains. These clumps can turn into primordia at a later stage. Fluctuating temperatures

Fig.21.1 : Shitake fruiting.

and a high CO_2 level promote knock formation. If many clumps are formed decrease the CO_2 level by slitting the plastic open.

(d) **Mycelial colouration:** Some aeration should be provided when the clumps have formed. The mycelium will turn reddish-brown..

(e) **Hardening of the coat:** Remove the plastic when bags have partially turned brown. The outside of the substrate (coat) will have gradually become hard, while the inside should be softer and moister. The brown hard skin acts like the bark in wood log production. It protects against contaminants and keeps the humidity in the substrate.

Fruiting

Fruiting indicated at the end of complete growth.

Harvesting

Hold the mushrooms by their stalks and break them off carefully from the substrate. Do not tear them from the surface otherwise too much substrate will be torn loose. Harvest the mushrooms at an early stage according to the quality requested by the buyers. Do not water the scars left behind for three or four days. White mycelium growing on the scar is a sign of recovery. Completely opened mushrooms have a much lower value in Asia, whereas buyers in Europe are less critical. Normal yields are 15 to 35% of the wet substrate weight.

Chapter 22

CULTIVATION OF WOOD EAR MUSHROOMS

Wood ear mushrooms (*Auricularia spp.*) are commonly cultivated in Asia. There are many Auricularia species of which *Auricularia polytricha*, *Auricularia fuscosuccinea* and *Auricularia auriculu-judea* are the most commonly grown. *Auricularia polytricha* is the most suitable species to cultivate in tropical regions where temperatures are high. Plastic bag cultivation is gaining popularity due to the scarcity of suitable logs and the ease with which different species of *Auricularia* can be cultivated on sawdust.

Taxonamic position

Kingdom : Fungi
Division : Bacidiomycota
Class : Agaricomycetes
Order : Auriculariales
Family : Auriculariaceae
Genus : *Auricularia*
Species : *Auricularia polytricha*

Cultivation

Substrate preparation

Sawdust of any one locally available tree is taken at about 80 kg and it is mixed with wheat bran (19 kg) and calcium carbonate (1 kg). Water content should be adjusted to 60-65% and pH to be adjusted to 5.5-6.0 using gypsum. Saw dust is soaked in water for 16-18 hours and wheat bran for three hours. All the ingredients are thoroughly mixed.

Filling and sterilization of bags: Fill the bags (1.5 to 2 kg) immediately after mixing all the ingredients. Otherwise fermentation and contamination may start. Polypropylene (heat resistant) bags are used for filling. The bags are first loosely filled and later pressed to get cylindrical shape. After filling the bag PVC or iron ring is inserted at the mouth of the bag and plugged with non-absorbent cotton. Sterilization is carried out in an autoclave at 22 psi for $1^1/_2$-2 hours.

Spawning and Spawn running: Spawning is carried out by removing the cotton plugs. Grain spawn is introduced @ 3% (dry weight basis) under aseptic conditions. After inoculation bags are placed in cropping rooms where these are incubated in a 4 h/ 20 h light/ dark cycles at 22-26°C. Spawn run may take 60-80 days During the period it goes through mycelial growth, mycelail coat, mycelial bump, pigmentation/browning and coat hardening phase.

Mycelial coat formation: A thick mycelial sheet coat will develop on the surface of the substrate. This will be formed after 6- 8 weeks of inoculation/ spawning.

72

Mycelial bump formation: Bumps are clumps of mycelium, commonly formed on the surface of most strains after 9-10 weeks. These bumps can turn into mushroom primordia at a later stage but most of them abort. Fluctuating temperatures and high CO_2 promotes bump formation.

Fruiting: The optimum fruiting temperature of wood ear mushrooms *(Auricularia polytricha)* is 23°–28°C. To promote primordia formation, the cotton plugs should be removed from the bags and holes cut in the bottom. Cuts in the bags are made so the mushrooms can emerge. Take care when handling the bags, because the texture of the substrate will stay soft even after the mycelium has colonised it. The mycelium is very sensitive to breakage. Only little light should be present in the mushroom house. Try to keep the temperature below 30°C by spraying water and opening the mushroom house at night. The primordia will develop into fruiting bodies in seven to ten days. Twist the fruiting bodies from the substrate by hand, leaving no bits of stem behind.Three to four flushes can be expected. Per bag of 1.2 kg, 300–500 g can be harvested.

Chapter 23

CULTIVATION OF GANODERMA LUCIDUM

It is a medicinal mushroom. It is also called reshi. The mushroom was found infrequently in nature. This lack of availability was largely responsible for the mushroom being so highly cherished and expensive. Current global trade of about 2 billion dollers; trade in India has crossed Rs. 100 crores annually. Reishi is reported to possess anticancer, antiHIV, antiheart attack (cholesterol lowering as well as anti-angiogenic), Hepato- and nephrotoprotective, hypoglycemic (anti-diabetes), antioxidants etc. Recently, the National Research Centre for Mushroom made a breakthrough in developing the cultivation technology of Reishi.

Taxonamic position

Kingdom	:	Fungi
Division	:	Bacidiomycota
Class	:	Agaricomycetes
Order	:	Polyporales
Family	:	Gonodermataceae
Genus	:	*Gonoderma*
Species	:	*G. lucidum*

Method of production

Reishi can be grown by the farmers seasonally in the low cost growing rooms preferably polyhouses. This mushroom is exclusively as medicine, it has to be grown organically. Reishi is grown on the saw dust of the broad-leaved trees (mango, poplar, coconut, sheesham). Sawdust is amended with 20% wheat bran and is wetted to a level of 65% moisture. Calcium sulphate (gypsum) and calcium carbonate (Chalk powder) are added to get a pH of 5.5. The mixed substrate (700 g dry wt; 2.1 kg wet) is filled in polypropylene bags the mouth of which is then plugged with cotton after putting a plastic ring. The bags are then sterilized in autoclave at 22 p.s.i. for 2 hrs. The substrate is spawned with wheat grain or saw dust spawn @ 3% on the dry weight basis. Spawn run (incubation) is done at 28-35°C in the closed rooms (high carbon dioxide) and darkness. After the complete spawn run (bags white all over), which takes about 25 days, polythene top is cut at the level of the substrate totally exposing the top side and proper conditions for fruiting or pinning (temp. 28°C, 1500 ppm CO_2, 800 lux light, 95% RH) are provided. Once the pins have grown up enough to form the cap which is indicated by the flattenning of the whitish top of the pinhead, humidity is reduced to 80% RH and more fresh air is introduced (1000 ppm CO_2). Once the cap is fully formed, which is indicated by yellowing of the cap margin (which is otherwise white), temperature is lowered to 25°C and RH is further reduced to 60% for cap thickening, reddening and maturation of the

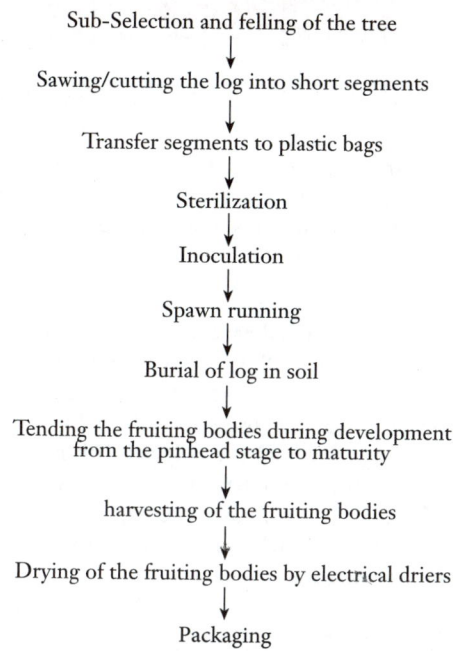

Sub-Selection and felling of the tree

↓

Sawing/cutting the log into short segments

↓

Transfer segments to plastic bags

↓

Sterilization

↓

Inoculation

↓

Spawn running

↓

Burial of log in soil

↓

Tending the fruiting bodies during development
from the pinhead stage to maturity

↓

harvesting of the fruiting bodies

↓

Drying of the fruiting bodies by electrical driers

↓

Packaging

Fig.20.1 : Flow chart of gonoderma cultivation.

fruitbodies. Full maturity is indicated, when the cap is fully reddish brown and spores are shed on the top of the cap. Harvesting is done by the tight plucking, holding the root with one hand and pulling up with another; scissors and knives can also be used but no residual bud is left after harvesting. One cycle of the growing takes 10-15 days. After harvesting the first flush, conditions for pinning are again switched on (i.e. 28 °C, 95%RH, 1500 ppm CO_2, 800 lux light) for staring and completing the second flush. Depending upon the conditions, 2-3 flushes appear and a total 25% B.E. can be achieved (250 g fresh mushroom from one kg dry substrate). One crop takes about four months. Harvested mushrooms, after washing with water, are dried at low temperature.

Chapter 24

CULTIVATION OF AGROCYBE AEGERITA

Agrocybe aegerita (Brig.) Sing., commonly known as 'black poplar mushroom'. It has good scope in India. *Agrocybe aegerita* is one of the tastiest mushrooms grown in temperate climates. It has unique flavour, good nutritive and medicinal values. This mushroom is known to have an antitumour lectin.

Method of cultivation

Wheat straw supplemented with wheat bran is commonly used as substrates for cultivation. Wheat straw is wetted thoroughly with water for 16-18 hrs. After wetting 5-10 per cent wheat bran is added in the saw dust and mixed thoroughly. Polypropylene bags are used for the cultivation. Two kg wet substrate is filled in each bag. The bags are plugged with non absorbant cotton by inserting a ring in the mouth of the bag. The filled bags are sterilized in the autoclaves for 11/2-2 hour at 22 p.s.i. After the bags have been sterilized and cooled down to room temperature, they are inoculated with 2-4% wheat grain based spawn. The Bags are placed/arranged in incubation rooms. The optimum temperature for the mycelial growth is between 25 and 28°C. Mycelia spread over the whole bag after 25-30 days.

Once the mycelium has fully colonized the substrate and formed thick mycelial mat and is ready for fruiting. At this stage the bags should be opened. Good fruit bodies are encouraged to form by adjusting the humidity in the room and by correct moisture content of the substrate. Developing fruit bodies Crop ready to harvest Small primordia start appearing after 5-8 days after opening the bags which become ready to harvest in the next four days. Average weight of a single fruit body is 3.5g.The fruit bodies could be sun dried or could be stored in the refrigerator for 7-10 days. On an average 300g of fresh fruit bodies can be harvested from half Kg dry wheat straw, thereby giving 60 percent biological efficiency.

Chapter 25

CULTIVATION TECHNOLOGY OF MILKY MUSHROOM

Calocybe is commonly called as Dhuth chatta. In India it is called milky white mushroom. The first ever milky white mushroom variety *Calocybe indica* P&C var. APK2 was released from Tamil Nadu Agricultural University, Coimbatore, India during 1998. Its milky white color and robust nature are appealing to consumers . Milky mushroom has been standardized on locally available substrates. This is the first Indian mushroom variety to be commercialized by IIHR. This variety is a tropical species and has excellent shelf life. Small scale mushroom growers prefer to grow this tropical mushroom due to the following reasons: (1) ideally suited to warm humid climate (30~38°C; 80% to 85% humidity), (2) its longer shelf life without any refrigeration (can be stored up to 7 days at room temperature), (3) retains fresh look and does not turn brown or dark black like that of button mushrooms, (4) lesser contamination due to competitor molds and insects during crop production under controlled conditions, (5) infrastructure needed to grow this mushroom is very much affordable and cost of production is comparatively low, which means industrial production could be attractive, and (6) has a short crop cycle (7~8 wk) and good biological efficiency of 140%. Milky mushroom (*Calocybe indica*) is second tropical mushroom after paddy straw mushroom, suitable for cultivation in tropical and subtropical regions of the country.

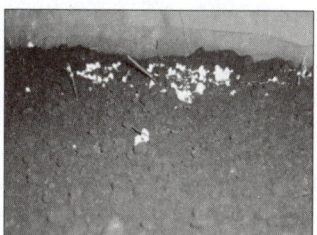

Fig.25.1 : Pin head.   **Fig.25.2 : Matured mushroom.**

The flow chart of the technology is as follows.

Substrate preparation (Paddy or wheat straw)

↓

Chopped in small (1-2cm pieces), soaked in water for 2-3 hours

↓

Substrate can be pasteurized by hot water (80°C) for 2 hours

↓

Hot water pasteurization or steam pasteurization

↓

Fill one Kg pasteurized substrate (65% moisture) in Polypropylene bags,
plug with non-absorbent cotton

↓

Spawning (Aseptically, @ 5% of wet substrate)

↓

Spawning in pasteurized bags

↓

Spawn running in dark room (30-38°C, 25-30 days)

↓

Casing with a mixture of red soil or red soil+ Decomposed coir compost
(1:1), pH 8.0
Shift o cropping room after casing, 30-38°C, humidity 80-85%
light, ventilation

↓

Pinhead initiation (7-10 days after casing)

↓

Harvesting (3-4 days after pinhead initiation, 30-38°C,
humidity 80-85%, light, ventilation)

↓

Sold as fresh, dry mushroom or as mushroom products

Cultivation process

Milky mushroom (Calocybe indica) can be grown on wide range of
substrates. It can be grown on substrates containing lignin, cellulose and
hemicelluloses. Substrate should be fresh and dry. Substrates exposed to rain or
harvested premature (green colour) are prone to various weed moulds which
may result in failure of the crop.

Straw is chopped in small pieces (2-4 cm size) and soaked in fresh water for
8-16 hours.

Water is boiled in wide mouth container and chopped wet straw filled in
gunny bag is submersed in hot water for 40 minutes at 80-90°C to achieve
pasteurization. This is very popular method particularly with small growers.

Substrate is filled in polypropylene bags (35 x 45cm, holding 2-3 kg wet
substrate) and sterilized at 15 lb psi for 1 hour. Once pasteurization/sterilization
is over straw is shifted to spawning room for cooling, bag filling and spawning.

Higher spawn dose 4-5% of wet substrate is used. Layer spawning is
preferred. After spawning bags are shifted to spawn running room and kept in

dark where temperature 25-35°C and relative humidity above 80% are maintained. It takes about 20 days when substrate is fully colonised and bags are ready for casing. Spawn run bags

The casing means covering the top surface of bags after spawn run is over, with sterilized casing material in thickness of about 2-3 cm. Casing material is spread in uniform layer of 2-3 cm thickness and sprayed with solution of carbendazim and formaldehyde to saturation level. Temperature 30-35°C and R.H. 80-90% are maintained.

Croping takes about 10 days for mycelium to reach on top of casing layer when fresh air is introduced while maintaining temperature and R.H. as above. Light should be provided for long time (10-12 hour daily). The changes thus made in environment, result in the initiation of fruiting bodies with in 3-5 days in the form of needle shape which mature in about a week. Mushrooms 7-8 cm diam. are harvested by twisting, cleaned and packed in perforated polythene/polypropylene bags for marketing. Mushrooms can also be wrapped in klin film for longer storage.

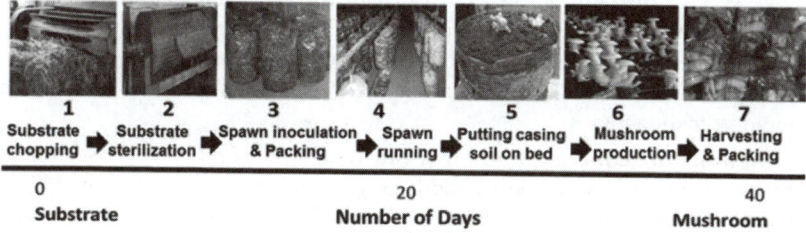

1	2	3	4	5	6	7
Substrate chopping	Substrate sterilization	Spawn inoculation & Packing	Spawn running	Putting casing soil on bed	Mushroom production	Harvesting & Packing

0	20	40
Substrate	Number of Days	Mushroom

Fig.25.3 : Steps in Mushroom Cultivation.

Chapter 26

FACTORS INFLUENCING MUSHROOM GROWTH

Commercially successful cultivation of mushroom depends on many factors such as temperature, moisture content of the substrate, pH, air circulation, relative humidity, quality of the substrate and spawn etc. These factors play an important role for the mycelial growth and fruiting body development.

(a) **Temperature** : The optimum temperature for mycelial and vegetative growth of white button mushroom is 22 to 25° C. During its vegetative growth phase temperature above 28° C results in poor mycelium growth and above 35° C is lethal to mushroom mycelium. After 10 days of spawn running, the temperature being to rise due to very active spawns growth, so precaution should be taken to check rising temperature by introducing fresh air and water.

In oyster mushroom, the required temperature is 25-28° C during spawn running. Rising of temperature should be checked through water. Temperature above 35 °C causes lethal effect on mycelial growth. The temperature required for the paddy straw mushroom is around 30° C. The temperature required for reproductive stage of oyster mushroom is 20-30° C. The paddy straw mushroom develops well at 30-35° C, but it can tolerate upto 45° C. Optimum temperature should be maintained.

(b) **Moisture content** : Moisture content of the substrate is very essential for active growth of mushrooms both vegetative and reproductive phases. Mushroom crop should be protected from dry atmosphere. The moisture content of the substrate should be 70%. The surface of the substrate should be kept moist by covering the substrate with paper and on paper gentle washing is to be done. Direct application of water on bed is more injurious to growing mycelium. Normally, substrates in polythene bags dose not require additional watering because of very low moisture loss, but in beds, watering depending on the atmospheric condition of mushroom house. At this stage no ventilation is required to maintain moist atmosphere.

For white button mushroom casing layer requires fairly wet condition. The casing materials should always be moist throughout its depth. Casing material should not be allowed to dry as small pin heads may fail to develop. On other hand, increased ventilation is required to prevent a thick layer of whitish tissue forming on the surface of the casing. Pinheads will not form when stroma has appeared. Excess of watering is harmful to the crop. Watering should be avoided during pinheads initiation as it leads to the damage of pinheads. Regular watering is required when the pinheads reached the size if 5mm diameter. The frequency of watering depends on atmospheric conditions, quality of casing material and ventilation of the room.

(c) **Relative Humidity (Rh)** : Mushrooms require high relative humidity during spawn running. The Rh should be 90 – 95 % inside spawn room. But during cropping of mushroom a slightly low relative humidity is required, ie.,

80 – 90%. During cropping period, high humidity (more than 80-90%) results in the bacterial infection, where as lower than optimum Rh results reduction in yield.

(d) **Ventilation** : Ventilation is required for maintaining suitable environmental condition and removal of toxic gas by introduction of adequate fresh air. During spawn running minimum ventilation is required which is well enough to change the air from the room. 4% CO_2 is required for good spawn growth.

Fresh air should be introduced only when it require to control the rise in temperature. Ventilation should be increased after spawn run. Low concentration of CO_2 (0.2%) during cropping produces abnormalities in mushroom crop and if it is too high, mushroom appears with too thick stalk and two small cap. For good crop of mushroom, the CO_2 concentration should be 0.7 to 0.9% (700-900ppm). The air speed should not increases 0.2 to 0.4m/sec. High speed of air causes scale formation and cracks in caps. If ventilation increases, Rh of the room is decreases.

(e) **Good Spawn** : Good quality disease and contamination free spawn produces higher yield. Young spawns are used for the cultivation, if aged spawn are used for spawning it takes larger time for developing mycelial growth and later fruit development. The spawn should be prepared with suitable substrate. Otherwise the spawn can not develop properly. For the good cultivation of mushroom, spawn must be of strain orienting from a single specimen or of a perfect crop. Substrate must be uniformly covered by the white mycelium. Appearance of any dark green, brown and black patches in mushroom indicates contamination.

(f) **Good Substrate :** Compost is the substrate for the cultivation of button mushroom. Synthesis of microbial proteins makes the fibrous material suitable for absorption and retention of the moisture. In the process of composting, physical and chemical changes of the substrate that is resist the growth of other competitive microorganisms. pH of the compost and substrate is about 7-8 and it should not have any amino acids.

Chapter 27

NUTRITIVE VALUE OF MUSHROOMS

The mushrooms are consumed by many people for their aroma and taste. As food, the nutritional value of mushroom lies between meat and vegetables. One hundred to two hundred gram of mushroom/dry weight is required to maintain nutritional balance in normal human beings with 70Kg body weight. Mushrooms are considered as alternative source of food. Mushroom products have a generalized or tonic effect, which in some cases may act prophylactically by increasing resistance to disease in humans from the balancing of nutrients in the diet and the enhancing of the immune systems. Mushrooms have been recognized by FAO as food contributing to the protein nutrient of the developing countries which are developing largely on cereals. Mushrooms are suited to supplement diet that rich with protein. The digestability of a food product does not necessarily bear any correlation to its nutritional value. Instead, its appearance, taste, and aroma, sometimes can stimulate one's appetite (preference). In addition to nutritional value, mushrooms have some unique colour, taste, aroma, and texture characteristics, which attract their consumption by humans.

Protein and amino acids

Protein is building material for all body tissues and is made up of various aminoacids. Mushrooms contains 25-35% Protein on dry weight basis. Mushroom protein contains all the nine essential amino acids required by man. The moisture content of fresh mushrooms varies within the range of 70-95% depending upon the harvest time and environmental conditions, whereas it is about 10-13% in dried mushrooms. The protein content of edible mushrooms in general, is about twice that of onion and cabbage and four times and 12 times those of oranges and apples respectively. In comparison, the protein content of common meat is as follows: pork, 9-16%; beef, 12-20%; chicken, 18-20%; fish, 18 -20%; and milk, 2.9- 3.3%.

Moisture content

Fresh mushrooms contains 90% moisture whereas air dried mushroom contains 15% moisture. Moisture content of the mushroom is affected by environmental factors, such as, temperature and relative humidity during its growth and storage.

Carbohydrates and fibers

The carbohydrate content of mushrooms are quite low. Mushrooms lack any starch based materials but it contains manitol (0.95%), reducing sugar (0.28%), glycogen (0.59%) and hemicellulose (0.91%). The type of carbohydrate present in the mushrooms are trehalose, glycose, manitol and glucon. Young mushroom contains fructose, glucose and sucrose. The mushroom cell wall contains glycogen and chitin polymers. The Pleurotus sp contain 4.2% soluble carbohydrate, 1.66% pentosans and 32.2% hexosans. Fresh mushrooms contain relatively large quantity of fibers (3.32%).

Fats

Mushrooms contain very low level of fats. The crude fat includes representatives of all classes of lipid compounds. Mushrooms are rich in linoleic acid which is an essential fatty acid. Some mushrooms contain more fats including free fatty acids, mono, di and triglycerides. In Pleurotus sp. contain oleic acid, palmitic acid and linoeic acids. Some mushrooms contain arachidonic acid also. It has also been reported that a total lipid content varying between 0.6 and 3.1 % of the dry weight, is found in the commonly cultivated mushrooms. At least 72 % of the total fatty acids are found to be unsaturated in all the four tested mushrooms. It should be noted that unsaturated fatty acids are essential and significant in our diet and to our health.

Vitamins and minerals

Mushrooms are a relatively good source vitamin and minerals. Phosphorus, iron, thiamine, riboflavin, ascorbic acid, ergosterine and niacin are available in plenty. It is also agood source of vitamin B1 and B2.

Vitamins and mineral content of *Agaricus bisporus*

Composition	Percentage
Thiamin	8.9
Riboflavin	3.7
Niacin	42.5
Ascorbic acid	21
Ca	71
P	9.2
Fe	8.8
Na	100
K	2850

(a) **Nutritive value of white button mushroom** : White button mushroom is a substance of high nutritive value. It contains a high percentage of protein but comparatively less animal protein. Besides, it contains appreciable amount of mineral matters and vitamins which are important for human body.

Composition	Percentage
Moisture	90.4
Crude protein	88
Fat	3.1
Carbohydrate	60
N-free content	51.1
Fiber	8.3
Ash	9.4
Energy (K cal.)	3.5

The carbohydrate of *A. bisporus* contains 4.2% soluble carbohydrate, 1.66% pentosans and 32.26% hexosans on dry weight basis. Since mushrooms are devoid of starch it is to be stored in the form of glycogen. The mushroom contains 2.8% fat. the crude fat is representative of all classes of lipid compound including **b) Nutritive value of *Pleurotus* sp.**

Pleurotus species are highly nutritious, palatable delicious for all age group of person. It is high in protein in compression to other vegetable proteins, but it has lesser protein than animal. It also contains some vitamins and minerals like calcium, phosphorous, iron, potassium and copper which are essential for human health. It is a good nutritious food for diabetic patients as it contains very low quantity of carbohydrates. The proximate composition and minerals are given in the table.

Table 27.1 : Mineral content of Pleurotus species (mg/100g dry weight)

Species	Iron	Calcium	Phosphorous	Potassium
P. Sajor-caju	120	35	1427	2680
P. flabulatus	122	34	1264	2790

Source : Memuna Haque (1989)

Table 27.2 : Aminoacid composition of Pleurotus species

Amino acid	P. sajor-caju	P. flabulatus
Aspartic acid	8.2	7.7
Threonine	5	4.3
Serine	5.8	4.9
Glutamic acid	14.1	9.2
Proline	5.2	4.4
Glycine	7.7	6.9
Alanine	8.6	8.0
Valine	5.2	5.9
Methionine	3.1	4.1
Iso-leucine	4.7	5.0
Lucine	9.7	7.5
Tryosine	2.5	3.4
Phenyl alanine	3.1	4.8
Histidine	2.3	3.8
Lysine	6.8	6.6
Ammonia	7.7	13.8]
Arginine	4.3	3.2

Source : Memuna Haque (1989)

(c) **Nutritive value of Paddy straw mushroom :** Paddy straw mushroom is nutritious palatable and delicious food for all types of people. It contains a high percentage of protein which is higher than any other vegetable protein but less than, animal protein. It also contains some vitamins and mineral like calcium, phosphorous, iron, potassium and copper which are essential for formation of bones and teeth. Mushroom consumption can help to overcome lysine deficiency as mushrooms contain lysine in high concentration. It has been observed that mushroom proteins are similar to those produced by animals which make then especially useful as supplement to cereal. The shortage to protein can be compensated by growing of mushroom on large scale in developing countries.

Table 27.3 : **Proximate composition of paddy straw mushroom (Volvariella species)**

Mushroom	Moisture	Ash	Protein	Fat	Fiber
V. diplasia	90.4	1.1	3.9	0.25	1.67
V. volvacea	88.4	1.46	4.90	0.75	1.38

Source : Zalia (1976)

Table 27.4 : **Vitamin and Mineral content of Volvariella species (mg/100g dry weight basis)**

Mushroom	Thiamine	Riboflavin	Nia-cin	Ascorbic acid	Ca	P	Fe	Na	K
V. diplasia	1.2	3.3	91.9	20.2	71	677	17.1	374	3455
V. volvacea	–	–	–	–	58	1042	17.7	-	3333

Chapter 28

POST-HARVEST HANDLING

Mushrooms are perishable food. After harvest they often change in ways that make them unacceptable for human consumption. The most readily observable of these changes include browning, liquefaction, loss of moisture, and loss of texture, aroma wilting, ripening, and flavour. Ensure that mushrooms are acceptable and nutritious to the consumer at the time of purchase. Expansion of the pileus by growth of gills and elongation of the stipe post harvest is supported by increased cell wall chitin and protein. The rate of cap opening depends on the stipe length (the longer the stipe, the greater the expansion), indicating that the stipe is acting as a major nutrient source for the expanding gill tissue.

Technologies such as cooling and modified atmosphere packaging can be use to delay the rate of senescence. Preservative technologies such as canning, drying, pickling, and freezing and-irradiation arrest biological function to prevent self damage. The shelf life of mushrooms may vary from one day to two weeks. Fresh mushrooms are best to stored. It is important that fresh mushrooms are packaged in materials that allow them to breathe, so they do not 'sweat' and become slimy. At the same time, the material should ensure mushrooms do not dry out too much.

Methods of post harvest handling

(a) Picking the mushroom

It is very important that the edges of the mushrooms are still curled under when they are picked. When the edges flatten, the spore will be released into the air. The release of spores does two things, it makes the air dangerous for workers, it can result in severe hay fever or asthma when workers breathe. Tills of freshly picked *Agaricus* mushrooms stacked on a cart, to be taken to packing area. Oyster mushrooms are more fragile them. Spores represent a significant weight, so there is less to sell. Errors are often made in picking, so without damaging the sporocarp the mushroom should be picked from the bed. An additional reason to pick early is that after the spores leave, the mushroom has no biological purpose.

(b) Trimming

The mushrooms have been picked, they have a little substrate attached. The substrate must be cut away, have some stem. Stems are generally difficult to chew, so it is not favoured by customers. The stems are removed by trimming. The stems add weight and if customers do not understand that they are getting a better product, it may be wise to leave the stems. Stems can be used to make other products, but that requires extra equipment.

(c) Steaming

Very similar to blanching in liquid, except steam instead of liquid is used to blanch.

Method

1. A steamer basket is placed in a boiling water bath such that it remains above the level of the boiling liquid. Mushrooms are steam blanched 2 minutes or so.

2. The mushrooms are then cooled either in a cold water bath and then drained and packed in freezer containers, or the mushrooms are spread directly from the steamer basket onto a cookie sheet sprayed with Pam or vegetable non-stick, spread into a single layer and then frozen on the cookie sheet until fully frozen.

3. They can then be freed from the cookie sheet and placed into freezer bags as individually free mushroom pieces.

4. These can be used in soups, stews, etc., or can be placed frozen directly into an ongoing stir-fry for great results.

Advantages

1. Convenient, more versatility than water blanching.

2. Excellent taste and texture preservation.

Disadvantages

1. Somewhat more time-consuming and more complicated than water blanch.

2. Does not clean dirt, sand, and grit as well as water blanching, so pre-selecting the cleanest mushrooms are best one.

(c) Freezing

Method

1. ice, chop, or prepare mushroom shape/pieces as preferred.

2. Fry in butter or olive or walnut oil for standard fried mushroom dish.

3. Stopping the cooking process slightly before normal and allow to cool by transferring mushrooms to cool pie plates, glass or baking dishes, etc.

4. When cool, portion mushrooms into small freezer containers and freeze.

5. To use, simply pop out the portion onto a sauté pan with a little of the same oil or butter used to first prepare them.

Advantage

1. Best reproduces the texture and taste of a mushroom.

2. Easy and convenient.

Disadvantages

1. Usually more air exposure inside freezer container; it may not preserve the mushroom in good quality.

2. Patting mushrooms gently to bottom of container into solid block and then placing some plastic wrap directly on mushrooms before covering with container lid helps some.

(e) Drying

The age old method of drying mushrooms is still one of the best methods of preserving mushrooms. Many good home use dehydrators are on the market.

Tunnel drier construction

A tunnel drier consists of a blower to circulate air, a heater to increase the temperature of the air to approximately 40 to 50°C (104 to 122°F), a place to put the food to be dried mushroom.

Method

1. Slice/prepare mushrooms, set on drying trays or rack and sun dry (protect from insects and flies) or place in dehydrator or oven (very low temperature 100°F to 150°F).

2. Dry to low moisture level, place in air tight bags or jars and store.

3. To re-constitute cover with warm or hot water 15 minutes to several hours (varies with species) until plumped up.

4. Do not use only the water from the tap for your recipes; be sure to use the soak water from the mushrooms as it is richly flavored.

5. Save it for other later uses if the recipe does not call for liquid. Freeze it irXno immediate us is at hand.

Advantages

1. Drying preserves mushrooms for very long periods of time with little or no deterioration in flavor or quality.

2. Drying actually intensifies the mushroom flavor of many species, especially the Boletes.

3. Convenient and easy to store and use, requiring no special equipment or refrigeration.

Disadvantages

1. Drying often toughens or changes mushroom texture; many will not fry or sauté after being dried.

2. Sometimes flavour changes character after drying occurs.

3. Volatile flavors and aromas are often lost.

4. Re-constituting does not necessarily restore good texture in many cases.

5. Equipment can be elaborate and expensive.

(f) Canning

Since mushrooms have insufficient acid, they are susceptible to *Clostridium botulinum* (botulin) contamination and require pressure canning to be safely canned.

Advantages

1. Versatility of product.

2. If properly done, anything can be canned, so stews, soups, preparations containing mushrooms can be prepared, then canned.

Disadvantages

1. Expensive, sometimes finicky equipment is necessary, complicated processes and preparation possible and very strict adherence to methods, procedures.

2. This is not a method to where one can cut corners.

(g) Salting

It is a preservation method in its own right or can be used to induce lactic fermentation. It works and ferments the mushrooms well, but it yields a very salty product. Less salt risks spoilage before fermentation begins. Salting to "dry" mushrooms is a technique similar to salting of fish. The water is drawn out by the salt and allowed to drain of or evaporate, leaving behind preserved product. Soaking re-constitutes product but often requires multiple rinses, and as mushroom flavor is lost this way, it is better to air dry mushrooms if moderate quantities or more are to be used. But for small amounts to bring up salt levels of a dish, this remains a very good method.

Advantages

1. Preservation stable and can actually change the cooking qualities less than full drying does, so some versatility exists here for texture with cooking.

2. Easy to do and stable end product.

Disadvantages

1. Mainly in high salt concentration forcing use of smaller quantities, or repeated rinses which leaches flavor needlessly.

2. A first quick, but thorough rinse to remove salt before the mushrooms absorb much water helps to minimize this problem.

(h) Packaging

It has become almost a universal practice to pack mushrooms in plastic or paper trays and to over wrap the trays with plastic film. The film protects them from the hands of customers and holds in moisture. Mushrooms are still alive as long as they continue to look good. The film will also restrict the oxygen that the living mushrooms require, so we must be certain that there is some place was oxygen can enter the package. Over wrapped trays have been used for more than 30 years for *Agaricus*, but it has now become almost universal for all kinds of mushrooms. Packing rooms should be clean and comfortable for workers.

Chapter 29

MUSHROOM RECIPES

Mushrooms are popular as a delicacy and have good nutritive value. It is a well established fact that they are excellent source of vitamins and minerals. A fresh mushroom contains about 85% to 95% moisture, 3% proteins, 4% carbohydrates, 0.3-0.4% fat and 1% minerals. Mushrooms are good source of thiamine, riboflavin, niacin and ascorbic acid. Being low in sodium content they are useful in adding flavour to dishes served without salt. During cooking there is loss of vitamin B in the juice, but in soup the vitamins are remain intact. With low carbohydrates and fat contents, they constitute an ideal diet for diabetic patient. Mushrooms have found a prominent place in Indian food system. Stems are very tasty and have a variety of uses like preparation of sauce, soup and vegetables. Example sliced or whole part of mushrooms used for making white sauce, soupe etc. Larger size may used as salad or vegetables. The rich flavour of open mushroom makes ideal complement too many other dishes.

Important highlights for mushroom recipes

- Mushrooms are purely vegetarian diet.
- Edible mushrooms should be identified and differentiated from other non-edible poisonous ones.
- Mushrooms should be purchased fresh, should not have a decaying rotten smell and slimy appearance.
- Before cooking, mushrooms should be cleaned with water to remove all adhering debris.
- Peeling or scrapping the outer skin of the mushroom is not advisable as it also results in loss of nutrients.
- Mushrooms can be blended with other ingredients as per the individual taste preference.

Some Indian and continental dishes of mushrooms are listed below;

1. Mushroom nonveg soup

Composition :

Mushroom - 100g; Meat stock - 250 ml; Onion - 50g; Tomato - 50g; Milk - 20 ml; Butter - 10 g; Maida Refined flour - One table spoon; Salt and pepper to taste.

Preparation :

1. Boil mushroom and meat stock for 15–20 minutes.

2. Heat butter and fry refined flour slightly then add onion and tomatoes.

1. Add milk and cook for 2 minutes and add cooked mushroom along with meat stock.

2. Add salt and pepper and serve hot.

2. Mushroom vegetarian soup

Composition :
Mushroom – 200g; butter – 25g; celery rip – 8g; onion – 500g; salt – 4g
white peper – 4g; milk – 400ml.

Preparation :
1. Fry the cleaned mushroom in melted butter and allow the liquid to
evaporate, add sliced anion and cool until the vegetable is soft.

2. Transfer the mixture in the jar of electric mixy and add full amount of
milk, salt and pepper and blend this mixture for 30 min.

3. The mixture is transferred to sauce pan and is heated for 5 minutes.

3. Mushroom prawn cutlets

Composition :
Mushroom - 50g; Prawn - 50g; Tomato - 100g; Onion - 100g; Coriander
leaves-one small bunch; Potato - 200g; Ginger   Small piece; Garlic-5
or 6 pods; Haldi powder - one teaspoon; Salt to taste; Green chillies - 6 to 7;
Rawa - 100g; Ghee for frying - 100g.

Preparation :
1. Shell the prawns and keep aside. Boil the potatoes and peel it and chop
the onion and tomatoes. Grind coriander leaves, ginger, garlic and green chilies
into a paste.

2. Mash the potatoes and slice the mushrooms.

3. In a vessel, melt the ghee and fry the onion, green masala and tomatoes.

4. Fry till onion and tomatoes are soft. Add prawns, mushrooms and mashed
potatoes.

5. Add salt, haldi powder to it.

6. Make the above mixture into balls; roll it in rawa and shallow fry in ghee
on tawa.

4. Mushroom fried rice

Composition :
Rice - 100g; Mushroom - 50g; Green peas - 50 g; Carrot - 50g;
Capsicum - 100g; Ajinomotto powder – Pinch; Lemon – 1; Bay leaves - 2 nos;
Cinamon - 2 to 4 pieces; Clove - 5 to 6; Pepper - 7 to 8; Oil - 80 ml ; Spring
onion - 25g.

Preparation :
1. Boil the rice till it is partially cooked then adds salt, bay leaf and drain.

2. Slice the mushrooms, carrots and capsicum.

3. Boil green peas and carrots.

4. In a vessel, heat oil and fry cloves, cinnamon, pepper then add capsicum
fry for sometime then add remaining vegetables and ajinomotto powder and
fry till vegetables are soft.

5. Add rice to the cooked vegetable, oil and lemon juice. Cook for 2 minutes.
Serve hot.

5. Sweet and sour mushrooms

Composition :

Mushroom - 100g; Cauliflower - 25g; Pepper powder - 1 teaspoon (tsp); Carrot - 25g; Capsicum - 50g; French beans -10g; Onion - 25g; Maida - 10g; Cornflour - 20g; Tomato- 150g; Or tomato ketchup - 2 tablespoon; Oil for frying - 60 ml.

Preparation :

1. Clean and cut vegetables into cubes.

2. Make a batter with maida and cauliflower, pepper powder and salt.

3. Cook the tomatoes and make a puree.

4. Dip the mushrooms and other vegetables into batter and fry till golden brown.

5. In a vessel, fry onion and capsicum till it is soft.

6. Add tomato puree or ketchup, boil it then add fried vegetables and salt and simmer for sometime.

6. Mushroom Pizza

Composition :

Refined flour - 60g; Fresh yeast - 5g; Salt - ° teaspoon (tsp); Mushroom - 50g; Tomato -75g; Onion - 20g; Cheese - 20g; Butter - 1tsp; Garlic - 1 flake; Chilli powder - 1 pinch.

Preparation :

1. Sieve the flour with salt. Mix the yeast into flour and add enough water to make dough.

2. Cover the dough till it becomes double in size and knead the dough.

3. Roll out the dough to 1 cm thickness. Grease the pizza tray and place the rolled out dough into tray.

4. Chop the onion, tomatoes and mushrooms. Saute the onion, garlic, tomatoes, mushrooms, chilli powder and salt. Cook for 5 to 6 minutes.

5. Spread the mixture over the rolled dough and sprinkle with grated cheese. Bake in a very hot oven for 20 minutes.

7. Mushroom samosa

Composition :

Mushroom- 100g; maida - 500g; ballery onion - 300g; dalta - 50g; green chillies - 15g; cardamom - 5nos; cinnamon - 1 -2; garlic - 10 cloves; ginger - small piece; gingelly oil - 500ml; salt to taste.

Preparation :

1. Mix dalta with maida flour and prepare dough.

2. Cut the mushroom, onion, ginger, green chillies into small pieces.

3. Peel garlic and ginger and mash them along with cinnamon and cardamom.

4. Take small amount of gingelly oil or refined ground nut oil in a frying pan and fry the mushroom, onion, green chillies.

5. Add the mashed masala and salt and as the mix turns to a thick paste, remove the pan from the fire.

6. Roll the dough into poori shape and place the masala inside.

7. Cut the sides using samosa spoon or knife and fry and serve hot.

8. Mushroom bonda

Composition :

Potato - 250g; mushroom -250g; bengal gram dhal - 100g; groundnut oil - required quantity; gingelly oil - 250g; chilli powder - 1tsp; ginger - small piece; green chillies - 10nos; coriander leaves - 1no; salt to taste.

Preparation :

1. Boil potatoes, peel and mash them along with salt.

2. Grind green chillies, ginger and coriander leaves.

3. Cut mushroom into small pieces. Roast mushroom in the oil.

4. Mix smashed potatoes, masala, roasted mushroom, chilli powder, and salt and make small balls.

5. Make thick batter from Bengal gram dhall by adding water.

6. Dip the balls into the batter and fry it in the oil till it turns to golden brown colour

9. Mushroom curry

Composition :

Mushroom - 500g; asafetida - 5g; curry leaves - 50g; red gram dhall - 1tsp; black gram dhall - 1tsp; pepper - 1tsp; fenugreek - 1/2tsp; red chillies - 5nos; coconut kernel - ° cup; jaggery and salt to taste; tomato - 150g; mustard - 5g; coriander leaves - 50g; Bengal gram dhall - 1tsp; poppy seeds - 1/2tsp; turmeric powder - a pinch; gingelly oil to required amount.

Preparation :

1. Wash mushroom and tomato and cut it into pieces.

2. Add oil to the pan and roast red gram dhall, black gram dhall and Bengal gram dhall till golden brown.

3. Roast poppy seeds separately. Prepare milk from coconut scrapings.

4. Extract juice from the tamarind after soaking in water. Roaste fenugreek, chillies and coriander till brown.

5. Powder cinnamon, poppy seeds and other pulses.

6. Add oil to pan and heat and put tomato and mushroom pieces to it and add turmeric powder salt, tamarind juice, jaggery and powdered spices and allow boiling.

7. Add coconut milk to the gravy and few more minutes. Season the curry with mustard and curry leaves.

8. Garnish and serve hot.

10. Mushroom paneer

Composition :

Mushroom - 200g; Bellary onion - 3nos; Cumin - $^1/_2$ tsp; Coriander powder - 2 tsp; Tomatoes – 2 nos; salt to taste; paneer - 100g; Garlic- 8 cloves; Turmeric $^1/_2$ tsp; Garam masala - $^1/_2$ tsp; ghee- 2 tsp.

Preparation :

1. Wash and cut mushrooms.

2. Warm ghee, add cumin to crackele.

3. Fry ground garlic, chopped onions and ginger till it becomes golden brown.

4. Add other species followed by chopped tomatoes and allow to siuncer.

5. Then add fried paneer and mushrooms and Cook for 10 minutes.

11. Mushroom kuruma

Composition :

Mushroom - 100g; Small onion - 50g; Green chillies - 20g; Grated coconut - $^1/_2$ cup; Pepper - 1tsp; Coriander powder - 1tsp; Roasted Bengal gram dhall - 1tsp; Cinnamon, bay leaf required quantity; Turmeric powder - $^1/_2$ tsp; Bellary onion - 200g; Tomato - 40g; Garlic - 5 cloves; oil - 75ml; Ginger- small piece; cloves - 1-2; salt to taste.

Preparation :

1. Clean and chop mushroom and fry with little oil and cut keep aside and small onions and green chillies.

2. Slice the tomatoes into small cubes.

3. Fry the above ingredients along with cinnamon, cloves and bay leaves, and grated coconut.

4. Make a paste with garlic, pepper, ginger and the above fried ingredients with required amount of water.

5. Cook this in water with salt after adding mushrooms and tomatoes and continue cooling for 5-10 minutes.

12. Mushroom vegetable pickle

Composition :

Mushroom - 1kg; Turnip - 1kg; Green peas - 1 kg; Ginger - 100g; Garlic - 20g; Acetic acid - 5tsp; salt - to taste; yellow radish - 1kg; Cauliflower - 1kg; Jaggery - 1kg; Black gram - 100g; Chilli powder - 100g; Gingelly oil - 1.5 kg.

Preparation :

1. Cut the mushroom in to pieces as required.

2. Heat oil in the frying pan and fry the mushroom.

3. Cut radish, turnip and cauliflower into small pieces and boil along with peas in water for 3-5 minutes.

4. Prepare syrup using jaggery with water and add acetic acid followed by the mushroom, cooked vegetables and fried garglic, ginger and stir well and remove from the fire.

5. Roast black gram powder and spread the powder on the prepared pickle and stir well.

6. Fill this in the glass bottle and keep in the sun for 3 days.

13. Mushroom soup

Composition :

Mushroom - 100g̈; Ginger - 20g; Corn flour - 2-3 tsp; Butter - 2 tsp; Pepper powder to taste; water - 1.5 lit; Small onion - 5g; Garlic - 10g; Milk - 2-3 cups; salt required amount; Green chillies to taste.

Preparation :

1. Clean and chop and peel ginger and onions.

2. Heat butter, add the corn flour and fry till it turns golden brown.

3. Add the milk and cook the mushroom with ginger, garlic and onions in1.5 lit water.

4. Collect the excess water by straining.

5. Mix the above ingredients and cook for 5-10 minutes.

6. Add to this required amount of pepper and salt.

14. Marinated mushrooms

Composition :

Mushroom – 200g; Green chillies – 6 nos; garlic – 6 cloves; gram dhall flour – 2 tsp; ghee for frying; vinegar – 7-10ml; salt to taste; curd – 1 cub and red colour – trace.

Preparation :

1. Add salt to whole mushroom and leave for 1-2 hr to remove excess water.

2. Dip in a mixture of curd, vinegar and colour and store in refrigerator for 48 hrs.

3. Prepare a thin paste of garm dhall flour.

4. Add ground garlic and chopped chillies.

5. Dip mushrooms in to this mixture and deep fry.

15. Stuffed mushroom

Composition :

Mushroom – 400g; bellary onion – 15g; fresh cream – 60g; chopped coriander leaves – 15 g; butter – 5g; salt and pepper to taste.

Preparation :

1. Heat the butter and fry the onion for $^1/_2$ minutes.

2. Add the prepared mushrooms and fry for mix well one minute.

3. Add the cream, chopped coriander leaves, salt and pepper and mix well.

16. Mushroom biriyani

Composition :

Mushroom – 100g; cinnamon – 5g; cardamom – 2 nos; cloves – 5 no; oil – 1tsb; rice – 200g; bellary onion -2 nos; tomato – 1 no; bay leaves – 4 nos and salt to taste.

Preparation :

1. Grind cardamom, cloves and cinnamon coarsely and tie in a piece of muslin cloth and boil in a pot of water.

2. Remove the muslin cloth bag from the water.

3. Chop onion finely and keep it separately.

4. Wash and cut mushrooms and boil in the above pot of water for 10 minutes and then strain the mushrooms and keep aside.

5. Wash rice and boil the same in the stock, when it is half cooked, remove the same in another pot.

6. Heat oil in another pan adds bay leaves, chopped onion and mushrooms and fry till it turns brown. Add salt to taste.

7. Add half cooked rice and steam for 3-6 minutes with the left over stock serve hot.

17. Mushroom omelet

Composition :

Butter - 3tsp; Mushroom sliced - 50g; Eggs – 4; Cream - 10g; Salt and pepper to taste.

Preparation :

1. Melt 1 teaspoon butter in a pan.

2. Sauté mushroom till soft. Beat eggs, cream, salt and pepper.

3. Melt 1 teaspoon butter in a pan.

4. Pour in eggs; cook till under side of omelet is cooked.

5. Put mushroom in the center. Fold just enough to set the eggs and serve hot.

18. Mushroom Bhajia

Composition :

Mushroom - 100g; Onion - 50g; Green chilies - 5-6 ; Besan - 150g ; Coriander leaves - $^1/_2$ bunch (small); Jeera - 1 teaspoon (tsp); Haldi powder - $^1/_2$ teaspoon (tsp); Chili powder - $^1/_2$ teaspoon (tsp); Oil for frying - 80 ml; Salt to taste.

Preparation :

1. Chop the onion and slice the mushroom.

2. Mix all the ingredients and make batter for bhajia.

3. Fry in oil.

19. Mushroom cutlet

Composition :

Mushroom - 100g; Potatoes - 250g; Bellary onion - 100g; Peas- 50g; Green chillies - 2 nos;   Garlic -10g; Ginger -small piece; Garam Masala - $^1/_2$ tsp; Turmeric powder - $^1/_2$ tsp; Chilli powder salt - to taste; Powdered coconut - 1tsp; Maida or egg white - 2 tsp; Bread crumbs - 3tsp; Oil for required and few Coriander leaves.

Preparation :

1. Cut the cleaned mushrooms and potatoes. Boil them with the soaked peas.

2. Mash the cooked potatoes. Add the chopped onion, green chillies, ginger, garlic, coriander leaves, garam masala, turmeric powder, chilli powder and powdered coconut.

3. Add salt to taste and mix them with coocked mushrooms and peas and make in to small balls. Make a better with maida and add salt to taste.

4. After flattening the base into the batter roll in bread crumbs and fry it in oil.

Chapter 30

MUSHROOM DISEASES

The mushroom diseases can be caused by both fungi and bacteria. Once the disease is introduced in the farm, secondary infection can be carried out by different agents such as air, water, machines, unhygienic equipments and handlers. The important fungal, bacterial and viral diseases and their causative organisms, symptom and control measures are described below.

A. Bacterial Diseases

1. Bacterial Blotch

It is a common bacterial disease of button mushroom. It is generally called bacterial spot. It was reported in 1915 from America and 1976 from India. The disease is common in mushroom farms where there is no healthy crops and poor ventilation system. It is a dreaded disease and may cause total loss of the crop. It has been reported in *Agaricus bitorquis* mushroom and *P. gladioli pv. Agaricoli*.

Causative agent : *Pseudomonas fluorescens*

Symptoms : The pathogen produces deep pale yellow colouration on the mushroom, later the colour changes to golden yellow or chocolate brown. Its occurrence is noticed from early buttons stage onwards and it also common on stored mushrooms. When pin heads are attacked they turn completely brown and do not develop into wild mushroom. Under favourable moisture conditions spots enlarge and covering the entire mushroom caps. Stipe will also show the symptom. Affected mushroom become sticky. Splitting at the blotched area is also noticed.

Method of transmission

1. Casing material and air-borne dust are the primary sources of infection.

2. It is present in the casing soil even after pasteurization.

3. Bacteria on the mushroom cap will reproduce easily.

4. High relative humidity and low temperature of 8 to 18oC favour the infection.

5. The bacterium survives on the mushroom spores, surfaces in debris, peat, chalk and on tools used in mushroom production.

6. Secondary spread is through hands of pickers, tools, ladder, implements, debris and mites.

Control

1. Casing materials before and after mixing should be stored in areas free from the pathogen.

2. Diseased mushrooms should be removed and destroyed.

3. Preventive measures should be taken to check spread through picker's hand and watering.

4. Adequate hygienic measures reducing the humidity (Rh should not exceed 85%) in the room (to introduce sufficient fresh air into cropping room).

5. Spraying with chlorinated water (100-150 ppm, 0.5 liters /m2) at three or four day's interval starting before the disease appearance.

6. Spraying with streptomycin 200 ppm or oxytetracycline 300 ppm is effective in controlling the disease.

2. Bacterial rot

The bacterium, *Pseudomonas alcaligens* is the causative agent of bacterial rot in *Pleurotus sajor-caju*. The symptoms include water-soaked spots and yellowish brown discolouration of young sporophores and rotting of matured sporophores. Rotting starts from the center of the sporophore towards periphery. The gills on the lower surface turn yellow. The cap become crinkled and rolls upward and inward.

3. Brown spot

Pseudomonas stutzeri is reported as competitive bacterium on paddy straw substrate used for the cultivation of *Pleurotus sajor-caju*. It induced brown spots in the substrate and caused 25 to 60 per cent yield reduction. Dipping in streptocycline solution (100 ppm) or formalin (25 ppm) controlled this bacterium.

4. Yellow bloch

Yellow blotch on *P. sajor-caju* is reported to be caused by *Pseudomonas agarici* and it caused complete crop failure in Solon, H.P. in India. The disease appears as blotches of varying sizes on pileups. The blotches are depressed, yellow and hazel-brown or orange in color. If the disease is noticed during primordial stage the entire crop will be lost. The infected fruit bodies become rot, shiny and emit bad odor at higher temperature and humid conditions. Spraying with ox tetracycline (400 ppm), streptocycline (400 ppm) or sodium hypochlorite (400 ppm) effectively controls the pathogen.

B. Virus Disease

A serious infectious disease of button mushroom was observed in the USA in 1948 and it was known as La France disease. Gandy and Holings (1962) demonstrated the association of three types of viruses in mushroom having die-back symptoms. In India, virus-like diseases were reported in white button mushroom (Tewari and Singh, 1984, 1985).

Causative agent: Spherical virus particles of 24 to 26 mm dia meter have been reported in *Pleurotus ostreatus, P. sapidus, P.columbinus* and *P. florida* and flexuous rods of 40-600 nm long. A polyhedral virus measuring 34 mm in dia has been reported on **Volvariella volvacea**. Rod and spherical viruses have been reported on *Lentinus edodes*.

Symptoms: The symptomatic expression of virus diseases on mushroom is due to the concentration of the virus strain, mushroom spawn used and environmental conditions given during cultivation. Due to virus infection the mycelium disappears after the surface of the casing layer. Sporosphores have off-white caps and mature early. Caps may be small and flat. Stipes are slightly bent elongated and watery in nature. Diseased mushrooms are loosely attached to substrate.

The gills become hard. Sometimes diseased mushroom give out musty smell.

Method of transmission

1. The viruses spread through mycelium, spores, and germ tubes of mushrooms and through vectors.

2. Phorid flies and mites help in introduction of virus particles to trays free virus infection.

Control

1. *Agaricus bitorquis* has been reported to be immune to all viruses affecting *A. bisporus*.

C. Fungal Diseases

There are four important fungal diseases of the cultivated mushrooms, particularly they infect to *Agaricus bisporus* and these are:

1. Wet Bubble

Wet bubble is also called as mycogone disease, white mould and bubble disease. Wet bubble was first reported from Jammu and Kashmir India in 1978. It is a serious disease of white button mushroom.

Causative agent: *Mycogone pernicious*, mycelium is white, compact and felt-like. Hyphae are hyaline, branched, septate and measure 3 to 5 μm in width. Conidiophores are hyaline, short, slender, branched and measure 200 x 3 to 5 μm in size. They have sub-verticillate branches. Conidia are thin-walled and one-celled. They measure 5 to 10 x 4 to 5 μm is size. Chlamydospores are large, spherical and two-celled. Its upper cell is thick-walled, globose, bright coloured and measure 15 to 30 x 10 to 20 μm in size. Lower cell is hyaline, smooth and measure 5 to 10 x 4 to 5 μm in size.

Symptoms: When the young pin heads are infected, the whole mass of mushroom becomes distorted and brown as scleroderma mass of white and fluffy. Due to the formation of chlamydospores it becomes brown and then decay. When the beds are infected the disease appears in spots. White mycelial patches occur on the surface of casing following infection of a developing mushroom below the casing surface. When the infection occurs at a later stage of mushroom development, brown stalks are formed on the stalk and gills. The affected gill shows white mycelial growth.

Method of transmission

1. The infection comes from casing soil,, air on the surface of the building, crop debris and spent compost.

2. Chlamydospores survive in casing soil upto three years.

3. The fungal conidia produced on the infected mushroom spread by air, water splashes, flies, mites and by pickers.

The optimum temperature for the mycelial growth is 25oC and the fungus infects few wild fleshy fungi also.

Control

1. Adoption of strict hygienic measures reduce wet bubble incidence.

2. The casing soil should be sterilized properly.

3. Spraying with zineb 0.3 % or mancozeb 0.3 % at weekly interval controls the disease. Use of benomyl or chlorothalonil or thiabendazole is also recommended. If casing soil is found contaminated treating the soil with one per cent formation at 25 liters per square meter is effective. Immediately after casing, spraying with 0.8 % (2 litres in 100 litres of water for 100 m2) on the casing surface can also be followed.

4. Immediately after the spray the mushroom house should be kept closed for 8 to 10 hours. Later, ventilation should be provided to remove the formaldehyde vapour (Formalin spraying is harmful at larger quantity).

2. Dry Bubble

Dry bubble was first reported in India (1973). It is named as *Verlicillium* disease, brown spot and La mole. It delays the pin-head formation, reduces the number of sporophores and the yield of white button mushroom. If it is not controlled, it may destroy the crop in two to three weeks. It is more common when the cropping period is extended beyond 60 days.

Causative agent: Verticillium fungi cola var. The fungus produces numerous hyaline, single- celled, thin- walled, oblong to cylindrical conidia and measure 3.5 to 15.9 x 1.5 to 5.0 μm in size. Conidia are produced on lateral or terminal and vertically branched conidiophores. Conidiophores are 220 to 800 x 1.5 to 5.0 μm in size. Conidia accumulate in clusters surrounded by sticky mucilage.

Symptoms: The symptoms of the disease vary with the time of infection. The fungus produces white mycelia growth on the casing soil later it turns into grayish yellow. If the infection occurs at pin head stage the whole tissue is look like bubbles.

If the infection occurs at a later stage, the diseased mushrooms become crooked and deformed. Hare-lip symptom is seen when a part of the cap is affected. The fungus grows on the mature sporophore as grey mould and makes them as unusable. On matured mushrooms it produces light brown and sunken spots on the caps. The spots in the cap coalesce and the caps shrink, turn leathery, dry and cracking. The fungus produces dark brown lesions on mushrooms. Sometimes grey mycelium covers the lesion especially in the centre. Cracks are found in the lesions on the caps.

Method of transmission

1. The major source of contamination is debris and dust on the floor of the mushroom house farm by infected casting soil.

2. Conidia are produced in sticky clusters and sticks easily to any contact.

3. The mites, *Sarcophagus* spp. feed on the mycelium and conidia carry the fungus from one house to another and from one farm to the other.

4. It can spread to other growing rooms by spores in air, by phorid and sciurid flies, equipments and hands and clothing of workers.

5. Conidia are spread by splashing and running water. Excess water running off the beds carries the conidia to lower beds or to the floor of the mushroom house.

6. The fungus is soil-borne and conidia survive in most soil for one year. It also perpetuates as resting mycelium in the infected sporophore and spent compost.

7. High humidity (90-95%), lack of proper air circulation, delayed harvesting and temperature above 15-17oC favours the development and spread of disease.

Control

1. Use of sterilized casing soil, pasteurized compost and proper disposal of spent compost helps in reducing the incidence of the disease.

2. Affected patches may be sprayed with 2 % commercial formalin. Spraying with mancozeb (0.25 %) or zineb (0.25 %) at 10 days interval controls the disease.

3. Provision of ideal environmental conditions like Rh to 80 to 85 % and temperature to 14 oC helps to reduce the disease incidence.

3. Green moulds

Green moulds *Trichoderma viridae* was first reported in India on *Agaricus bisporus* (1977), later *T. hamatum* and *T. harzianum* were reported during 1986-1987. *T. hamatum* caused about 15 % loss in *A. bisporus* (1987). *T. harziamum,, T. koningu, T. lignorum* and *T. viride* have been reported on *Agaricus bisporus, Pleurotus* spp. *Pholiota maneko* and *Lenzites edodes*.

Causative agent: Different *Trichoderma* spp. is resposible for green mould disease on mushrooms. They are *Trichobermma hamatum. T. koningu* and *T. viride, T. hamatum*: Mycelium is hyaline, septate and branched. Conidiophores branches at right angles. Phialides are pin-shaped and arise singly or in whorls. Phialospores are dark green and measure 2.5 to 7.5 x 2.5 to 3.0 μm. Phialospores of *T. harziamun* are pale green and measure 4.0 to 6.5 x 3.5 μm. Phialospores of *T. viride* are single celled, green, thin walled and measure 2.5 to 5.0 x 2.4 μm. It produces chlamydospores in old cultures.

Symptoms: *Trichoderma* spp. are associated with green mould symptoms in compost, on casing soil, in spawn bottles and on grains after spawning. *Trichoderma viride* attacks the spawning tray and reduces the spawn run. It appears as green patches on the spawned and cased trays. It reduces the pin head formation of the mushroom. *T. viride* causes reddish brown discolouration of the stripe and sunken lesions on the pileus. *T. koningii* grows as cottony white or grey mycelium over the casing surface. It also produces purple brown spots with a dry cracked surface. Infected caps turn brown. On young pin heads enlarged spots occur leading to cracking of cap, later stipe is also infected. At high humid conditions it causes brown spots on the caps.

Mode of transmission

1. Addition of un-sterilized supplements introduces green moulds.

2. Dead mushroom tissues in the beds and cut stipes help in the infection.

3. Green moulds appear generally in composts rich in carbohydrate and poor in nitrogen.

4. High relative humidity and low pH of the casing soil flavor green moulds.

5. Air-borne dust and mites are main infection routes.

Control

1. Proper pasteurization and conditioning of compost reduces the green moulds.

2. Supplements should be sterilized properly before their use.

3. Dead mushrooms and cut stalks should be removed from mushroom house promptly and destroyed.

4. Proper hygienic conditions should be maintained during mushroom growing.

5. Spraying with zineb 0.2 % on the used soil control the green moulds. *T. lignorum* was killed by exposing soil 70oC for one hour.

6. *T. lignorum* and *T. koningii* were eliminated by holding the compost at 60oC for 12 hours and then at 40o C for five days.

4. Cobweb Disease

It is otherwise known as soft decay, dactylium disease and mildew disease. It causes soft rot or decay of fruiting body. It was first reported on white button mushroom (1977) from Himachal Pradesh. The disease caused 50% loss in yield.

Causative agent: *Clabotryum dendroides* or *Daetylium dendroides*, mycelium is hyaline, branched and septate. Conidiospores are erect, simple or branched. Conidia are single, elongated, 2 to 3 septate, slightly constricted at the base and measure 20 to 30 x 10 to 12.5 μm in size.

Symptoms: Small, circular, white patches of mycelium of the fungus appears on the surface of casting. Finaly the fungus grown as a fluffy white mould on the mushroom. The diseased mushroom turns brown and rot. The mycelium of the pathogenic fungus turns pink or red.

Mode of transmission

1. The pathogen is soil-borne and spread through contaminated soil.

2. In the farm spores are spread through air.

3. High humidity and temperature flavors cobweb disease.

Control

1. Sterilization of the casing mixture at 50o C for 4 hrs or disinfection of casing soil by binomial (150g/100m2 casing area) controls the disease.

2. The relative humidity and temperature during picking should not exceed 90 % and 65 oF.

3. Dusting between flushes with zineb or mancozeb at 00g/100m2 or spraying with formalin 0.2 to 0.3 % prevents the fungal attack.

5. Minor fungal infection

(i) *Cladobotryum* spp : *Clabobotryum apiculatrum, C. verticilliatum* and *C. variospermum* are reported in oyster mushroom. They are seen as white cottony growth on the substrate. The sporophore will show irregular sunken spots. Fluffy growth is seen on fruiting bodies. The diseased sporophores showed soft rot symptoms. Rotted sporophores emit bad odor. The fungi can be controlled by spraying with 50 ppm of carbendazim solution.

(ii) *Gliocladium* spp : *Gliocladium virens* and *G. deliquescens* grow on the sporophores of oyster mushroom and produce green spots. Attached young pin heads turn soft, brown, pale yellow and finally decay. Matured sporophores exhibit brown spots surrounded by yellow halo. These fungi can be controlled by spraying with 100 ppm solution of carbendazim.

(iii) *Sibirina fungicola* : The affected sporophore show powdery white growth on the stripe, gill and primordial. Later the primordial becomes soft. Matured fruit bodies become fragile. Provision of proper aeration and relative humidity in the mushroom house and spraying with benomyl control this pathogen.

Chapter 31

MUSHROOM ABNORMALITIES AND COMPETITOR MUSHROOMS

Mushroom growing out of another mushroom that appears to be the same species (abnormal, "freak" fruitings are not uncommon in the mushroom world, and do not actually represent mycotrophism. Mycotroph arising from a small mass of tissue (a sclerotium) or from copious white threads; stem of mycotroph about 1-2 mm wide; cap of mycotroph fairly smooth; gills of mycotroph well developed; spores of mycotroph produced only as basidiospores (on basidia located on the gills). Mycotroph not arising from a sclerotium or copious threads; stem of mycotroph wider than above (usually at least 5 mm wide when mature); cap of mycotroph smooth or becoming powdery; gills of mycotroph thick and distant, or poorly formed; spores of mycotroph produced primarily as asexual chlamydospores on the cap surface or gills. Cap of mycotroph becoming powdery with maturity; gills of mycotroph poorly formed from the beginning. Cap of mycotroph not becoming powdery; gills of mycotroph thick and well spaced, at least when young. Some of the common mushroom abnormalities are listed here.

1. False truffle

It is also known as calve's brain disease. False truffle was first reported from Solon, India in 1965. Its incidence is noticed bothin hills and in the plains of North India. First it was reported on A. *bitorqurs*. Severe crop losses occur when the crop is affected before, after and during spawn run. False truffle was reported from Uttar Pradesh, Himachal Pradesh, Punjab and Haryana in India. It occurs throughout the cropping period of hot weather conditions.

Fungal characters: Dichliomyces microsporus (Diel), *Pseudobalsamia microspora*. Ascocarps are fleshy. They contain many asci. Asci are oval or sub-spherical in shape with short or long stalks. Each ascus measures 19 to 27 x 10.5 to 15 μm are in size and contain three to eight ascospores. Ascospores are spherical, sulphur coloured with one distinct oil drop. Ascospore is 6.5 μm in diameter. It produces chlamydospores also.

Symptoms: The disease appears as small weft of white to cream mycelium on the surface of the compost. Later it becomes thick and develops into white, solid, wrinkled, round to irregular mass resembling brain or peeled walnut-like structure called ascocarp. Ascocarp appears in masses and raises the casing soil gently. Ascocarps are spherical to irregular, white to cream initially and turn brown at maturity. They finally disintegrate into a powdery mass omitting a chlorine-like smell. The fungus does not allow mushroom mycelium to grow. The infected mycelium turns pale brown. The spawn in affected spots turns soggy and disappears.

Method of transfer

1. Ascospores from casing soil and in wooden trays of previous crops are the sources of infection.

2. Spread of the ascospores occurs in drainage water and on air-borne debris.

3. Ascospores germinate at 30oC and its germination is stimulated by presence of actively growing mycelium of the mushroom.

4. Optimum growth of the fungus is 20 to 28oC.

Control

1. Strict hygienic measures should be followed.

2. Filter with 2 μm in diameter mesh should be installed in spawn running areas to prevent the fungus.

3. Compost should be prepared only on concrete floors.

4. Compost temperature during spawn run should not exceed 21 to 24°C.

5. Casing soil which contains ascospores should not be used. Young truffles can be picked and buried or burnt before fruit bodies turn brown.

6. Drying of woodworks and trays help eradication of the fungus.

7. Compost should be cooked out at 70°C for 12 hours or at 80°C for two hours at the crop to kill the ascospores.

8. Initial infection can be checked by treating the affected patches with 2 % formalin solution.

2. Olive green mould

Olive green mould (*Chaetomium olivaccum*) was first reported from India during 1975 and *C. glocosum* was reported in 1979. It reduces the yield ranging from 25 to 50 % in *A. bisporus*.

Fungal characters: *Chaetomium globosum* and *C. olivaceum* perithecia are superficial, opaque, globose, thin, and membranous with an apical tuft of dark bristles. Asci are clavate and evanescent. Ascopores are dark brown, broadly ovate, umbonate at both ends. They measure 9.0 to 12.5 x 7.0 to 9.5μm in size. *C. globosum*: Perithecia are scattered or gregarious and broadly ovate or ellipsoid. Often they are pointed at the base and thickly clothed with slender flexuous hairs. Asciare oblong to clavate. Ascospores are broadly ovate,measure about 8.0 to 9.5 x 6.0 to 8.0μm.

Symptoms: The fungus appears as grayish white mycelium in the compost soon after spawning. Spawn growth is delayed and reduced. Compost which does not support spawn will support the growth of *Chactomium* spp.

Method of transfer

1. Compost and casing soil are the major source of infection.

2. Ascospores are spread by air flows, clothes and other materials used in mushroom house.

3. The fungi are favoured by anacrobic conditions during peak heat which occur when the compost is too wet, over composted (temperature 62oC).

4. These fungi are able to survive at higher levels of ammonia.

5. The growth of C.olivaceum is favoured by alkalinity of the compost.

Control

1. Olive green moulod can be prevented by good composting practices.

2. It is effectively controlled by peak heat conditions.

3. Spraying with zineb 0.2 % controls the spread of the disease.

3. Brown plaster mould

Brown plaster mould caused by *Papulospora hyssina* was first reported from India in 1974. It causes 90 % *Agaricus bisporus*.

Fungal characters: *Papulospora byssina*, mycelium is septate and brown in colour. It produces clusters of brown coloured, many celled, spherical bubils measuring 60 to 130 x 30 to 190 μm.

Symptoms: The fungus appears as white mycelium on the surface of the compost and in casing soil. The colour changes to light brown to cinnamon brown and to rust colour at the end. In serve cases, the growth of the mushroom mycelium is completely arrested.

Method of transfer

1. The fungus spreads through air-borne bubils in mushroom house.

2. Highly moistured compost and improper pasteurization of compost, higher temperature during spawn run and cropping favour the development of the fungus.

3. The fungus is commonly seen in the compost, require quantity of gypsum is added to compost.

Control

1. Composting should be made with addition of recommended level of gypsum.

2. Peak heating should be sufficient duration and at proper temperature.

3. Compost should not be too wet.

4. Spraying with carbendazim 0.1 % benomyl 0.1 %, thiophanate methyl 0.1 % and carboxin 0.1 % are recommended for its control.

4. Yellow moulds

Yellow moulds were first reported on white button mushroom in Jammu and Kashmir (1978), and Punjab (1987), in India. It reduces 25 to 70 % yield.

Fungal characters: *Chrysosporium luteum* (Ces) *C. sulphureum* and *Myceliophthora lutea* (Cost) are the causative agents. Mycelium is white at first and turns yellow to dark, septate and branched. It produces three kinds of spores: a. Condition which are smooth, avoid and terminal b. Chlamydospores which are smooth and thick-walled and c. Chlamydospores, which are spiny and thick-walled.

Symptoms: These fungi form a yellowish brown corky mycelial layer at the interphase of compost and casing material. The yellow moulds may develop in a layer below the casing, form circular colonies in the compost (confetti) or they may be troubled throughout the compost. When it develops its stroma it becomes visible. They reduce the food supply to the mushroom or eradicate its mycelium by toxic metabolites.

Method of transfer

1. The primary infection is through chicken manure, spent compost and improperly sterilized wooden works.

2. The secondary spread is through flies, mites, water splashes, picking and tools used in mushroom growing.

3. Yellow moulds prefer the same conditions as the mushroom fungus.

4. Severity of yellow moulds increased in the compost with 70 % moisture and 19 to 20 oc temperature.

Control

1. Good farm hygiene and proper pasteurization of compost and casing layer.

2. Spraying with benomyl 0.04 to 0.05% or copper oxychloride 0.04% are also recommended.

3. Spraying with calcium hypochlorite solution 15% removes the moulds.

5. Speedonium yellow mould

Yellow mould disease caused by the fungus, *Sepedonium* was first reported in India at 1991. Sometimes it caused total failure of the crop.

Fungal morphology: *Sepedonium chryosporium* (Hypomyces chryospermum), mycelium is septate, hyaline and 3 to 5 μm in dia. Conidiophores are erect and bear lateral simple or botryose cluster of branches which are septate. At the tip of these branches conidia are borne singly. Conidia are hyaline, thin-walled, ellipsed or pisiform, produced singly from the tips of the phoalids. Chlamydospores are globose, wasted, dark yellow, thick-walled and 13 to 21 μm in diameter.

Symptoms: The mould occurs in the compost as white growth, finally the colour changes to yellow to tan. It occurs at the basal layers of compost or at the bottom of the cropping bags. The fungus causes deformation of mushrooms probably due to production of volatile toxins. These toxins also inhibit mycelia growth of the mushroom.

Method of transfer

1. Primary infection is from spent compost

2. Spores spread through air.

3. The fungus survives through thick-walled chlamydo-spores.

4. More wetness leads to its development in the lower layer of compost.

Control

1. Proper pasteurization of compost and provision of air-filters during spawn running (to prevent the entry of spores) reduces the incidence.

2. Incorporation of carbendazin 0.5 % in the compost effectively controls it.

6. Ink caps

Ink caps (*Coprmns* spp) appear generally during spawn can in North India on white button mushroom. It is a competitor mould.

Fungal morphology: *Copernicus armamentariums* and *C. funelarus* occur in white button mushroom beds. Caps are 1.5 to 4.0 cm wide and campanulate. Surface of the cap is white when the cap expands its margin splits. Gills are 6 to 10 cm long, free, white but soon turn black on liquefying. Stipe is 5 to 10 cm long, 2 to 3 mm in diameter, hollow white, glittery, delicate and tapers upwards with a small bulb at the base. Basidiospores are black, elliptic and measure 8 to 12 x 3 to 5 mm in size.

Symptoms: Caps appear in the compost during spawn run or in newly cased beds. The stipe of these weedy mushrooms is slender with bell shaped caps. Caps are cream coloured at first but later turn to bluish black. Sometimes it occurs in cluster. They decay and form a black shiny mass due to auto-digestion.

Method of transfer

1. The principal infection is through improperly pasteurized compost or casing soil.

2. Ink caps are seen when the compost has surplus nitrogen in the ammoniacal form, and when insufficient quantities of gypsum are added to the compost.

3. They may also occur when the compost is too wet.

Control

1. Proper pasteurization of compost soil, avoiding excessive watering and rouging out the weedy mushrooms from the beds.

2. Before filling the trays the compost should be free from ammonia.

3. If the fruiting bodies of the weedy fungus are formed in large numbers in spawned trays then the compost should be re-pasteurized at 60oC for two hours.

7. White plaster mould

In India the occurrence of white plaster mould has been reported from different parts (1989), and it reduces the yield about 35 % in white button mushroom yield.

Symptoms: The fungus appears as white dense patches on the compost or casing soil. The patches may be small to more than 50 cm in diameter later, the white growth changes to light pink. Spawn run is affected completely.

Method of transfer

1. Unpasteurized compost, manures and chilling of the beds cause occurrence of this fungus.

2. The fungus is favoured by over-composted compost with excess ammonia and has a pH of 8.2.

3. Excessive moisture and insufficient ventilation favour this fungus.

Control

1. Proper composting and addition of optimum quantities of gypsum and water reduce the fungus.

2. Spraying benomyl 10 % in formation 4 % solution reduce the incidence.

8. Lipstick mould

In India it has been reported from Punjab (1987) and Himachel Pradesh (1988). It is also called as red lipstick mould.

Symptoms: The mould appears as fine, cottony white mycelia growth in the cracks or crevices of casing soil or in the sides of compost trays. Matures fungus produces colour changes to cherry red and finally to dull orange or buff. The fungus inhibits mycelial growth of the mushroom.

Fungal morphology: *Sporendonema purpurascens* (Ben), mycelium is white, septate and segmented into chains of one-celled, short, red, cylindrical spores with shorten ends.

Method of transfer

1. Spent compost and casing soil are important sources of infection.

2. The chicken manure is suspected to carry this fungus.

3. The fungus is spread by pickers and water splashes.

Control

1. It has effectively controlled by heating with steam at 144o F for 90 minutes.

2. Apply calcium hypochlorite or benomyl or zineb.

3. Drenching the casing with zineb (1 Ib/450 litres of water at 4.5 litres/100 sq. ft of bed).

9. Pink mould

In India, pink mould has been reported in 1981. During white button mushroom cultivation it appears first as white growth on the casing soil and later it turns pink.

Fungal morphology: *Cephalothecum roseum*, mycelium is branched and septate. Spores are erect, branched and slightly swollen at the tip. Conidia are hyaline to pink, single, pear-shaped, two-celled. The apical cell of the conidium is larger. The conidia are spread through air.

Control

1. Pink mould is controlled by spraying twice at 10 days interval captain or thiram 0.04 % on the casing soil.

10. Fire mould

1. Fire mould, *Neurospora crassa* appears in the compost or casing after the cook out. It produces white mycelium which later turn orange.

2. Large wefts of mycelium hang down like cob webs.

3. It is very difficult to eradicate it once become established in the farm.

11. Smoky mould

1. *Penicillium* spp. associated with the cultivation of oyster and paddy straw mushrooms.

2. They are normally found growing on trays or side boards of shelves or pieces of fruiting bodies left on the bed.

3. They are also found on grain medium when there was improper sterilization during spawn preparation.

4. Affected beds smell moldy.

5. The fungus produces smoky spores in the air.

6. Provision of air filters in phase-II, strict hygienic measures and spraying with formalin solution controls the smoky moulds.

Chapter 32

INSECT AND PEST OF MUSHROOMS

Various species of flies are among the most dangerous mushroom pests. They are belongs to a group of insects called "dipterous" (double-wing) due to the fact that adult flies that are similar in size and colour. A correct identification of the pests is necessary for the selection of control methods depending on the cycle and the development properties of each species. They attracted by the smell of compost and growing mycelium. Flies get inside the growing rooms through every possible openings and cracks and lay eggs into the substrate, casing layer and on fruit bodies. At favourable conditions larvae appear from the eggs after 5-10 days and cause great damage to the mushrooms. Larvae feed on mushroom mycelium and gnaw the pins. Developing pins and young mushrooms are smaller than ripe mushrooms and therefore, they don't survive after the larvae attack. The damage of mycelium can also be very severe, which results in the formation of pins brown in colour and having a leathery surface and the appearance of holes in the stipes. In addition to the yield losses that larvae cause, feeding on mycelium, the mushrooms lose their marketable appearance because of the larvae that swarm on their surface. Larvae also enter fruit bodies forming many passageways and holes, which makes the mushrooms unsuitable for use. The important insects are described below.

1. Phorid file

In 1975 an unidentified phorid fly was a reported from India on white button mushroom *Megaselia agaric* Lintner, in1991 from Tamil Nadu in oyster mushroom *(Pleurotus citrinopileatus)*. The phorid fly, *M. agarica* is considered as one of the important pests in mushrooms.

Damages

1. The files are attracted to the mycelium odour in the spawned compost or the casing materials, tunnel throught the bottom of the stalk of the mushroom.

2. The larvae feed gregariously on the mushroom tissues and move upwards in to the cap or pileus, forming tunnels till their maturity.

3. Larvae also feed on decomposing spawned grains the damaged mushroom turns down along the tunnel in the stalk portion and the mushrooms become yellowish brown.

4. When the mushroom is attacked at the 'pin head' stage, its development is restricted.

Biology of the insect

1. The adult flies are hump-backed, light to dark brown in colour and measure 19 to 20 mm in length.

2. The abdomen of male fly is slightly curved downwards. In female fly the abdomen is pointed and straight at the end.

3. Eggs are white, 0.33 mm in length, elongated, cylindrical and slightly curved.

4. Fully grown larvae are dirty white, 3.3 mm long with transparent visible black 'mouth hooks'. Larvae are narrow at the anterior end.

5. The pupae are 2.27 mm long, dorsoventrally flattened, yellowish brown to dark brown. Fully grown pupae are having pair of black respiratory horns on the thorax.

6. Mating takes place outside mushroom house in open during evening hours in the air and flies comes back to mushroom house "in copula".

7. During rainy seasons flies breed on puff ball mushroom.

8. Males are active than females. Adult male and female flies survive for 2 to and 3 days respectively. From egg to adult emergence it takes 15 days. At 25°C the larval and pupal periods lasts for 5 and 7 days respectively.

Control

1. Disposal of spent up compost and casing materials

2. Provision of wire nets of 14-16 mesh/cm on doors, windows and ventilators check the entry and breeding of phorid flies.

3. Incorporation of Aldicarb at 100 ppm in the compost and casing is most effective in controlling larval population and infestation of mushroom followed by chlorphyriphos, lindane and diazinon.

4. At the final turning of compost 60ml of Lindane 20EC after dilution in 12 to 16 liters of water is thoroughly (spray) mixed with compost prepared from every 300 kg of wheat straw.

5. Dichlorvos is sprayed as low volume spray with spinning disc sprayer at 22.5 g a.i/100m3.

2. Sciarid files

Sciarid flies, *Bradysia pauper* (Toum) in 1975 and *B. tritict* (Coq) in 1980 from India were reported to damage white button mushrooms. In 1987, *Lycoriella auripila* (Winn) was reported on oyster mushroom in 1987. The yield loss due to sciarid flies in oyster mushroom ranges from 30.0 to 34.5%.

Damages

1. The larvae feed on thicken mycelia attachments and restrict the spawn run.

2. They feed on the mycelia attachments at pin head stage and cause severing and brown discolouration of pin heads.

3. The affected pin heads become leathery.

4. Larvae may enter into the pin heads and make them hollow. They tunnel into the stipes of mushroom ready for harvest, make them hollow and spongy and unfit for consumption.

Biology of insect

1. Adults are 2.2 to 3.2 mm long and grayish black in colour.

2. They have long thread-like antenna with 14 annuli in the flagellum. Fore wings have thick coastal margin formed by costa and sub-costa.

3. The abdomen of male fly terminals in claspers and the female the abdomen is swollen with pointed ovipositor. The female fly lays eggs mostly in spawned compost and casing materials.

4. A single female lays about 110 eggs in clusters. Sometimes they lay the eggs in chains or singly.

5. The eggs are pulpy white, round to oval and are 0.23 mm long.

6. They hatch in 7 days at 23oC.

7. Larva is dirty white in colour with distinct shiny black head and measure 6.1 to 7.3mm in length. Larvae feed on mycelium, mushrooms and decaying organic matter.

8. They become full growth in 8 or 9 days

9. The newly formed pupae are dirty white matured pupae are dark green to black and 2.29 to 2.95 mm in size.

10. Male pupae are smaller than female pupae. Pupal period last 4 or 5 days.

11. Matured pupa comes to the casing surface and then adult emerges. Adults live for 2 or 3 days at 22.50C.

12. Five generations are completed during the cropping period.

Control

1. Proper disposal of compost and casing materials.

2. The entry of the flies into the mushroom house can be checked by provision of nylon or wire nets of 14-16 meshes/cm in the doors and other inlets.

3. Incorporation of insecticide in the compost and casing soil reduce the larval population and thereby the damage.

4. At the final turning of compost (prepared from 300kg of wheat straw) incorporation by spraying of 60 ml of Lindane 20 EC after dilution in water effectively checks the larval population.

5. Diflubenzuron is applying to the compost and casing material at 50 ppm concentration effectively controls the larva.

3. Spring tails

Lepidocyrtus sp. and *Xenylla* sp. damages the beds of button mushroom have been reported in India in 1994.

Damages

1. The insect feeds on the mycelium of the white button and oyster mushrooms by scrapping it from spawn grains and wounding the mycelia strand.

2. This leads to striking of spawn run in white button mushroom.

3. Springtails feed and cause brown and shallow pits on white button mushrooms leading to its deterioration

4. This damage in oyster mushroom is comparatively more.

5. In young stage the upper and lower surfaces of the mushroom.

6. Spring-tails congregate near the base of the stalk and feed on the stalks and gills. The scrapped stalk turns brown in colour.

7. In paddy straw mushroom feeding by spring-tail causes small, irregular pits on cap and basal portion of stalk.

Biology of insect

1. Spring tails are wingless.

2. The adults of *Seira iricolor* are ground coloured with light violet band along the sides of the body.

3. The insect (including appendages) measures 2.85 mm in length.

4. The female insect lays eggs singly or in clusters on the moist paddy straw, compost and mushroom buttons.

5. Eggs are laid several times at an interval of 10 to 15 days.

6. Eggs are white, smooth and spherical in shape. Egg hatches in 3.2 days at 30°C Adult insects live upto 78 days at 26°C.

Control

1. The floor and inside of the mushroom houses and the surrounding area should be made free from any organic matter.

2. The spent compost and casing materials should be disposed off in far of places.

3. Effective pasteurization should be made to kill the springtails.

4. The composting yard floor, walls and surrounding areas should be disinfested by spraying malathion 0.05 % spray.

5. If the compost is infested by spring-tail, diazinon incorporated in the compost at 30 ppm concentration (15 ml of diazinon 20EC for 100kg after dilution in water) at the time of filling.

4. Beetle

The bettle, *Staphylinus* sp. was found damaging oyster mushroom and it was reported from Kerala in 1991.

Biology of insect

1. The beetles have short elytra and large membranous hind wings kept folded below the elytra when not in use.

2. The tip of the abdomen is held curled over the back.

3. The larvae are white long and campodeiform with tubular terminal segment. The adult beetles are attracted to the aroma from as well as discarded mushroom debris.

4. The grubs feed on soft gills of oyster mushroom.

5. They make small irregular holes in the htmenium and stipe in the initial stages and later on damage the developing mushrooms.

6. The life cycle is completed in about three weeks.

7. Adult beetle predate on mites and spring-tails.

Control

1. Debris of mushroom from the mushroom house and surrounding areas should be removed to prevent egg laying by females.

2. Over-matured mushrooms should be removed from the beds.

3. Application of bleaching powder in the mushroom house repels the beetles.

5. Nematodes

Nematodes damage mushroom crop and may causes total crop failure. First report of nematode damage in India, in 1972. *Mycophagous* (feeding on fungi) and *saprophagous* nematodes are very important in mushroom production. Nematodes occur at any time from composting to cropping. Composting materials (wheat straw, chicken manure, horse dung), casing material, platform, soil, water, trays, implements and spent composts are important sources of nematode contamination. The yield loss due to nematode infestation depends upon stage of cropping at the time of attack, species and inoculums. Total number of flushes is greatly reduced and the flushes are delayed by nematode attack. Infestation of *Aphenchiodes composticola* at spawning 10% loss in white button mushrooms. At the same level loss is 64 % when it infests at casing time. *Aphelenchoides sacchari* leads to total loss when in infests at spawning stage.

Damages

1. These nematodes feed upon the mycelia sap by piercing the hyphal wall with their stylet and devitalize the mycelium.

2. The important symptoms of nematode attack include sparse mycelia growth in patches, changing of mycelium to stingy nature, sinking of compost, browning of spawn run (instead of being white), delayed and poor mushroom flushes, alternate poor and high yields and reduction in number of flushes.

Control

1. It is better to avoid entry of nematodes rather than controlling them after their entry.

2. Thionazin at 80 ppm is recommended for nematode control in mushroom cultivation.

3. Incorporation of neem leaf powder (2) to 5 per cent (W/W), oil cakes of neem, Karanj, coconut or groundnut in the compost before spawning is recommended to reduce the population of nematodes and to increase the sporophore yield.

4. The compost is made free of nematode by maintaining the air and bed temperature in pasteurization room at 60°C for at least two hours and steam cook-out of mushroom house at 70°C for five to six hours or 80°C for 30 to 60 minutes.

Chapter 33

ECONOMICS OF MUSHROOM CULTIVATION

Since mushroom cultivation can be a labour-intensive agro-industrial activity, it could have great economic and social impact by generating income and employment for both women and youth, particularly in rural areas in developing countries. It is necessary to estimate the investment of the mushroom cultivation and the income to be earned. Example, in 1990 the total production of mushrooms in India was 10,000 tonnes, which accounted for less than 6 percent of total world mushroom production. In 2006, however, total production of mushrooms in India reached 50,000 tonnes and accounted for over 1.5 % of total world mushroom production. Total employment in the mushroom industry in India was over 2% increases in 2008, with only 10 percent of the employed being actual mushroom farmers, other employment fall within sectors such as food, beverage manufacturing, trading and management, transport, marketing, wholesaling, retailing, export etc.

Table 33.1 : Capital investment (for 50,000 bottles)

S. No	Name of the item	Price (Rs.)
1.	Room for autoclaving grain boiling and grain filling in bottles	35,000
2.	Laminar air flow	35,000
3.	Store room and Spawn multiplication room	25,000
4.	LPG/Electric stove	20,000
5.	Grain boiling container	500
6.	Hot plate	5,000
7.	Pressure cooker	1,500
8.	Others	2,500
9.	AC (one tons)	
	Total	**89,500**

The mushroom industry can also have even broader positive spill-over, generating complementary employment in areas such as accommodation, restaurant services etc. The local mushroom industry can also be the main source of revenue for local government.

The income obtained from mushroom cultivation also depends on many factors like demand, selling price, facility of its processing transport facilities and location etc. The management also affect the profit or loss of the farm. The spawn of good quality is essential for profitable production. If the farm near to the town where fresh mushrooms should b sold very easily this will give good return.

The mushroom can be grown throughout the year but this type of farm requires all facilities like temperature, humidity and air circulation control.

This type of farm require spawn production unit. It means autoclaving room, grain boiling and grain filling room, Laminar air flow, Store room and spawn multiplication room.

Table 33.2 : *Requiring expenditure (for 50,000 bottles)*

S. No	Particulars	Quantity	Amount
1.	Wheat/ or other grain	5000	50000
2.	Empty glucose bottle	5000	4000
3.	Non absorbent cotton	3000	10000
4.	Chalk powder	150	1000
5.	Gypsum	150	250
6.	Media and mother culture	1 Kg	1500
7.	Water and other		15000
8.	Labour		20000
9.	Other miscellaneous		5000
	Total		**2,00,200**

Table 33.3 : *Cost of production*

S. No	Particulars	Amount
1.	Recurring	1,01,750
2.	Fixed	89,500
3.	Depreciation on fixed investment @10%	8,950
	Total	**2,00,200**

Cost of production = 2,00,200

Cost of production per unit bottle = ~ 4/- Rs.

General rate of bottle in Rs. 15.00 profit per unit if sold. = Rs. 11/-

So, net profit = Rs. 5,50,000

Economics of a small scale model

The demand for button mushroom is fast increasing in international markets and a big gap exists between supply and demand. There is a need to take advantage of this situation by encouraging its production which is a highly viable venture as brought out below:

Table 17.4 : *Costs and Returns*

Project cost	Amount (Rs. in Lakhs)
Land and Site Development	5.15
Building	44.96
Plant and Machinery	47.00
Misc. Fixed Assets	0.75
Contingency	4.88
Pre-Operative Cost	4.25
Total	**106.99**

Table 33.5 : *Permanent item for non-recurring expenditure*

S. No	Name of the items	Nos.	Price Rs.
1.	Spawn staking and chopping shed 50' x 50'	1	
2.	Pasteurization room 30' x 17' x 12'	1	
3.	Bag filling room 30' x 17' x 12'	1	
4.	Spawn running room 30' x 17' x 12'	2	
5.	Cropping room 30' x 12' x 12'	5	
6.	Cleaning and packing room including office 30' x 17' x 12'	1	
7.	Storage room	1	
			4.50000
1.	Motor operator	1	8000
2.	Bag filling table and spawing	2	2000
3.	Racks	4	7000
4.	Sprayer	1	5000
5.	Boiling apparatus	1	5000
	Total		**4.77000**

Table 33.6 : *Working capital investment*

S. No	Name of the items	Nos.	Price Rs.
1.	Paddy straw	60 tons	60000
2.	Spawn bottles	30000	10000
3.	Chemical		1000
4.	Electricity and water		70000
5.	Polythene bags		1000
6.	Transport		65000
7.	Others		10000
	Total		**217000**

The major components of the model are:

1. **Cost of acquiring land and its development (Rs.5.15 lakhs):** The land would have to be acquired in areas well connected to urban markets. On an average the cost of land might be put at Rs.3 lakhs per acre.

2. Cost of leveling the site (including fencing etc.) would be Rs.0.15 lakhs and cost of putting up guard rooms would be Rs.50 thousand.

3. **Building (Rs.44.96 lakhs):** The estimated cost of this component works out to around Rs.45 lakhs, major item being growing room at the cost of Rs.25.92 lakhs.

4. Plant & Machinery (Rs.47.00 lakhs): The cost of equipping the production unit works out to Rs.30 lakhs, that of compost and casing unit to Rs.7 lakhs and that of installing canning facilities, spawn Lab and other equipments to Rs.10 lakhs.

5. Miscellaneous Fixed Assets (Rs.0.75 lakhs): This is the estimated cost of building up a communication system and furnishing.

6. Pre-operative Expenses (Rs.4.25 lakhs): These include professional charges, administrative expenses and other start up expenses.

Economics of oyster mushroom cultivation

Oyster mushroom cultivation does not require so many sophisticated equipments and standard permanent construction for its growing. It can be grown in temporary sheds. In this, polythene or polypropaline bags are used. The oyster can be grown throughout the year in humid and temperate regions (*Table 33.4, 33.6*).

Chapter 34

ENVIRONMENTAL IMPACT ON MUSHROOM CULTIVATION

Organic solid wastes are a kind of biomass, which are generated annually through the activities of the agricultural, forest and food processing industries. They consist mainly of three components: cellulose, hemicellulose and lignin. The general term for these organic wastes is lignocellulose. It is common knowledge that lignocellulosic wastes are available in abundance both in the rural and urban areas. They have insignificant or less commercial value and certainly no food value, at least in their original form. When carelessly disposed of in the surrounding environment by dumping or burning, these wastes are bound to lead to environmental pollution and consequently health hazards. It should be recognised that the wastes are resources out of place and their proper management and utilization would lead to further economic growth as well.

Reducing environmental pollution

In 1999, more than 3,000 million tons of cereal straws were available in the world, and about half of these residues remain unused. In addition, the world produced 952 million tons of bagasse; 6,476 thousand tons of coffee pulps; 6,152 thousand tons of coffee wastes; 9,386 thousand of cottonseed hulls; 14,073 thousand tons of sunflower seed hulls; and 325 thousand tons of sisal wastes. Million tons of sawdust, wood chips, and water hyacinth are also available worldwide. All these lignocellulosic waste residues can be used as substrate growing mushrooms; otherwise, they would cause health hazards.

Composting

Composting is the main process by which mushrooms can degrade and utilize the agro waste materials. Mushrooms are grown, not directly on soil as other crops, but on organic substrates like straw, corn cobs, saw dust bagasse, wood pulp, cotton waste, coconut huske, leaves etc either raw or composted. Used compost from mushroom growing can also be recycled for use as animal feeds soil conditioning and fertilizers. Composting of this agricultural waste consists of the piling up of substrates for a certain period of time during which various changes occur.

The purpose of composting is to produce a selective medium for the growth of mycelium. During composting certain physical and chemical changes occur that allow selective growth of the mushroom.

Not all mushrooms require composted substrates, Agaricus is generally grown on composted and pasteurized substrates where as Pleurotus and other species can be grown on various agricultural waste materials with the use of different technologies. Pleurotus species grow well on sifferent types of lignocellulosic materials, covering the materials in to digestible and protein rich substances suitable for animal feed.

Mushroom enzymes can break down lignin, cellulose and hemicellulose present in these organic materials into simpler molecules. Lignocellulosic compounds are complex and insoluble. They can be treated by various chemical methods, e.g. with dilute hydrochloric acid and calcium chloride to increase the digestibility and nutritional qualities, and even to form sugars to serve as carbon sources. However, these chemical methods are tedious and costly. Furthermore, treatments to eliminate adverse side effects of the chemicals are also very complex. In contrast, mushroom cultivation techniques have become significantly important in recent years in improving nutritional quality and upgrading the economic value of the solid organic wastes.

Mushrooms with other fungi are presently only organisms that can synthesize and excrete the relevant hydrolytic and oxidative enzymes that enable them to degrade complex organic substrates into soluble substances which can then be absorbed by the mushrooms for their nutrients, the ability of the different mushroom species to utilize various substrates will depend on both mushroom-and substrate-associated factors. For example, examination of the lignocellulolytic enzymes profiles of the three important commercially cultivated mushrooms exhibit varying abilities to utilize different lignocellulosics as growth substrate.

- *Lentinula edodes* is cultivated on highly lignified substrates such as wood or sawdust, produces two extracellular enzymes such as manganese peroxidase and laccase, which have been associated with lignin depolymerisation.

- *Volvariella volvacea* prefers high cellulose- low lignin-containing substrates such as paddy straw and cotton wastes which have relatively low lignin content, and produces a variety of cellulolytic enzymes including five different endoglucanases, five cellobi-hydrolases and two β-glucosidases, but no lignin-degrading enzymes.

- *Pleurotus sajor-caju* is the most adaptable species and can be grown on a wide variety of agricultural waste materials containing different chemical composition in terms of polysaccharide. Based on this it excrete both kinds of cellulose and lignin-degrading enzymes.

Organic Wastes Recycling

The ultimate purpose of the applied aspects of any scientific endeavour is to integrate wherever possible the various disciplines of science and technological processes in order that gain maximum benefits. Combined production of mushrooms, biogas and biofertilizer from the rural and urban organic wastes should be one of the aims of such integrated schemes that can eventually be put into profitable operation. Though the conventional and established approaches helps to produces food, fertilizer and fuel exist, the explosive growth of the population, the rapid depletion of conventional fuel resources leads mankind to look for alternative sources for food, fertilizer and fuel.

Even though man has been harvesting mushrooms as food from wild sources from times immemorial, their nutritive value was not assessed and their production under controlled conditions was not undertaken until recent decades. The lignocellulosic substrate used for mushroom production and which is left after harvesting of the mushrooms can be used as compost for soil conditioning. It should be noted that this compost besides being rich in nitrogenous material contains partly degraded lignocellulosic components, which when combined with animal dung or human excreta in a biogas digest would yield not only biogas but also a good quality organic nitrogenous fertilizer in the form of sludge. The sludge from the biogas plant as a nitrogenous fertilizer is far more beneficial than the compost from which it has been derived. Part of the biogas that is produced in the vicinity of the mushroom house can also be conveniently used for pasteurization of the mushroom bed material and maintenance of the optimal temperature in the mushroom house as well.

It is therefore suggested that an integrated approach in the production of mushroom, biofertilizer and biogas should be considered as a feasible approach for the rural and urban lignocellulosic waste utilization and disposal. This is the "Zero Emission or Total Productivity" concept.

Environmental restoration

Mushroom cultivation technology is friendly to the environment. Mushroom mycelia can produce a group of complex extracellular enzymes which can degrade and utilize the lignocellulosic wastes in order to reduce pollution. It has been revealed recently that mushroom mycelia can play a significant role in the restoration of damaged environments. Saprotrophic, endophytic, mycorrhizal, or even parasitic fungi/mushrooms can be used in mycorestoration, which can be performed in four different ways: mycofiltration (using mycelia to filter water), mycoforestry (using mycelia to restore forests), mycoremediation (using mycelia to eliminate toxic waste, and mycopesticides (using mycelia to control insect pests). These methods represent the potential to create the clean ecosystem, where no damage will be left after fungal implementation.

Chapter 35

MEDICINAL VALUES OF MUSHROOMS

Medicinal uses of mushrooms were known from ancient times. Greeks regarded mushrooms as "providing strength to soldiers in war". But Chinese regarded it as Elixir of life. Food and agriculture organization has recommended mushroom as a food to people in underdeveloped nations where protein malnutrition is a threat to human health. Earlier mushrooms were recommended for their medicinal values rather than for nutritional values. The growth and development of mushroom science have lead to us develop technologies for identifying, culturing and cultivation of several mushrooms in different countries in the world. The research of mushroom has brought out information's on the medicinal properties of mushroom from different countries. Mushrooms are low in sugar, starch and calories. They have more fibrous materials and contain easily digestible proteins. They are rich in vitamins and minerals essential for human beings. Ash and high fibre contents make mushrooms as best human diet for the peoples suffering from hyperacidities and constipation.

In Ayurveda, mushrooms are mentioned as body builders and energizers. Larger group of mushrooms, both edible and poisonous have medicinal properties and are used in specific diseases. At present, numbers of mushrooms have been reported to possess pharmacological values like antibacterial, antifungal and antiviral properties. Compounds isolated from *Flammulina velutipes, F. mellac, Clitocybe, Marasimius, Pleurotus, Psathy*rella, *Trichoderma* and *Agaricus bisporus* are characterized with antibacterial activity. Mushrooms have an antibiotic effect of human pathogens such as, *Aspergillus fumigatus* in *Cortinarius orellanus* and *Tricholoma ustaloides*. Spread of *Candida albicans* is prevented by extracts of *Tricholoma saponaceum*. The terpenoids, illudin-M and illudin-S extracted from *Clitocybe illudens* are very effective against *Plasmodium gallinoceum*.

In China, more than 107 edible varieties of mushrooms possess medicinal values. About 20 medicines are commercially prepared from mushrooms which include sedatives, anti cancerous, anti radiational drugs, liver protective, recuperating agents for stomach and intestine and medicines for stimulating bile secretion and for dizziness and headache. Chinese considered *Armillariella mellea, Coriolus versicolor, Gonoderma lucidum, A. tabescense, Herinium erinaccus, Lentinus edodes, Marasmius* and *Tremella fuciformis* as important medicinal mushrooms and they are using them in the form of extracts, powders (in capsules) and sugar coated tablets. The active principles in these mushrooms are said to be immunostimulating polysaccharides which strengthen health and immunity. Polysaccharides-peptides of *Coriolus versicolor* are anti cancerous while polysaccharide of shiitake is having liver protective activity.

The fungus, *Clitocybe illudens* has antiprotozoal effect. Illudin and other terpenoids from *Clitocybe illudens* is supposed to be active against a plasmodium.

A wide variety of mushrooms with antifungal activities include *Lentinus edodes*, *Cortinellus shiitake*, *Oudemansiella muci*dia and *Coprinus comatus*.

Among the edible fungi *Calvatia gigantean*, *Chlorophyllum molybdites* and *Agaicus campestris* have antiviral activities against human diseases. Among the non-edible mushrooms, *Russula emetica* and *Panacolus subaltealus* have antiviral principles, it induces the interferon formation. Extract of *Lentinus edodes* inhibits the growth of some virus like influenza.

Cultivated mushrooms such as white button mushroom (*Agaricus bisporus*), shiitake (*Lentinusedodes*) and oyster mushroom (*Pleurotus spp*) contain high amount of retina, a substance which may in some circumstances have an antagonistic effect on some forms of tumour. Research at National Cancer Research Institute, Tokyo showed that intraperitoneal injection of aqueous extracts of *Auricularia auricular*, *Flammulina velutipes*, *Lentinus edodes*, *Pholiota nameko*, *P. ostreatus*, *Pleurotus spodoeucus* and *Tricholoma matsutke* had some inhibitory effect on the growth of tumours.

An anti tumourous substance namely calvacin has been extracted from giant puffball (*Calvatia gigantean*). Similarly polysaccharides and basic proteins have been detected from *L. edodes* which are anti tumour medicines. A polysaccharide, lentinan from *L. edodes* is found anticancerous. Water soluble polysaccharide, fraction from *L. edodes* inhibited the growth of mouse sarcoma 180 in mice. Another polysaccharide KS-2 extracted from culture medium of *L. edodes* suppressed sarcoma 180 and Ehrlich ascites carcinoma.

* *Poria corticola* and *Boletus edulis* have anti tumour activity.
* The paddy straw mushroom *Volvariella volvacea* is a source for medicinal polysaccharide, a potential anti tumour compound.
* *Agaricus bisporus*, *A. blazer* etc. are also reported to have anti tumour principles.
* The mushroom *Ganoderma lucidum* contain lot of immune regulating compounds. It is called as longevity mushroom in Korea and Mushroom of Immunity in Japan.
* The maitake mushroom, *Gritola frondosa* has better anticancer property, recently it was reported that, it has an antitumour, anti HIV and helpful in reducing blood sugar, blood pressure and constipations.
* *Coprinus comatus* is observed to have an anti-diabetic effect. Regular consumption of this mushroom reduces blood sugar level.
* Regular consumption of *Tricholoma populinum* is believed to help against hayfever and allergic inflammation of blood vessels.
* Prominent mushrooms such as *A. bisporus* and *Lentinus edodes* lowering blood cholesterol and hypertension.
* *Auricularia polytricha*, *Flammulina velutipes* and *Fomes fomentarius* have less activity in reducing blood pressure.

- *Pleurotus sajor-caju* exhibited hypertention action and reduced the rate of nephron deterioration and chronic renal failure patients.

- *L.edodes reduses* plasma cholesterol both in animal and human beings and it induce the cholesterol extraction process from the tissues.

- *L. edodes* has anti hypolipidemic principles, it contains eritademine C2 ® dihydroxy-4- (9-adenyl butyric acid) and it affects cholesterol triglyceride and phospholipids levels.

- *Auricularia polytricha* possess platelet aggregation inhibitor named adenosine.

- *Volvariella volvacea* (Volvatoxin-A) and *Flammulina velutipes* (Flammutoxin) has cardio toxic activity, it lowers the blood pressure and are active against tumour cells.

Singapore Physician, Wu Shui has reported the pharmacological values of shiitake mushroom. This mushroom gives enhanced vigour, sexuality, energy and diminished aging. The Ajinomoto company of Japan has suggested that 'lentinan sulphate' of shiitake could inhibit the development of AIDS virus. Extract of this mushroom have used for clinical trials against AIDS.

They are considered to cure stomach, troubles and improve sexuality desire. Mushrooms boiled in milk with sugar is used for the treatment of tuberculosis.

In India especially from Kerala, 'Nilamanga' an underground mushroom is used for different diseases. This type of mushroom is used in different size, shapes and appearance and is usually associated with old termite nests under the soil. They appear singly or in bunch with small stalkes. Inner flesh of this mushroom is white and soft (young) and it become hard when it is matured. Boiled extract of this mushroom with sesame oil is poured in drops into the ear for curing ear ache for children's. Boiled milk extract of this mushroom is consumed as oral medicine in jaundice than any other medicine. Consumption of boiled water extract of Nilamanga prevents body dehydration.

- *Auricularia auricular* (Jew's ear) is frequently used as a poultice for inflamed eyes and throat.

- *Lycoperdon* giganteum is used as a soft surgical dressing.

- *Calvatia gigantean* is used as anesthesia.

- *Ganoderma lucidum* is a sedative.

- *Tremella fuciformis* is used as an anti radiational drug.

Aamanita muscaria (Fly agaric or fly mushroom) is a deadly poisonous mushroom. It is used in the form of powder and tincture for curing swollon glands, nervous troubling and epilepsy. It is used either externally or internally at higher dilution levels. It is a homeopathic medicine in the name of agar, agric and agarus. Two compounds such as ibotenic acid and musinol are prepared from this mushroom which is used to cure malfunction of GABA (Gama Amino Butyric acid) system of brain called Schizophore break or major inhibitory compound for central nervous system of human beings. It also used for the treatment of cholera and intermittent fever.

The sacred mushroom *Psilocybe mexicana* contain two chemicals like Psilocin and Psilocybin, used to cure mental disorders. The hallucinogenic property was due to psilocybin and psilocin. These chemicals are indole derivatives with similar effects. They stimulate the effect of LSD (d-lysergic acid diethyl amide) loss of sense of time, space and feeling extradinary lightness or hovering. *Fomes ignarius* and *F. fomentarius* are used for rapid coagulation of blood. *Polyporous officinalis* is considered as a home medicine. *Armilariella mellea* is an excellent purgative; extract of this mushroom is applied externally to stop bleeding. It is also used for chronic diseases of lungs and breast. It is a good medicine for night sweating in tuberculosis, rheumatism, jaundice, dropsy and intestinal worms. It is also used as a homeopathic medicine.

GLOSSARY

Actinomycetes: White filamentous organisms (sometimes similar to fungal hyphae), which occur in well-fermented compost, indicating that the compost is suitable for cultivation of *Agaricus* spp.

Acyanophilic: No colour change in the spore wall when treated with blue strain.

Adnate: Gills joined to the stipe at their entire width.

Agar: An extract from a seaweed used to solidify media: alternatively, (cheaper) gelatine may be used. Agar is available in bar or powder form.

Agaricus bisporus: common cultivated mushrooms, in a variety of lines.

Air lock: Enclosed section with gates at each end in order to prevent outside air entering directly the growing room.

Amatoxins: Cyclopeptides present in certain mushrooms. alpha-amanitin and beta-amanitin are the principal amatoxins.

Anaerobic fermentation: The reverse of aerobic or otherwise the lack of oxygen during fermentation - undesirable.

Anaerobic: Environment Without oxygen (O2).

Annulus: The ring found on the stem of certain species of mushrooms.

Anthracophilous: Inhabiting burned-over soil or growing directly on burned wood.

Antibiotic: A chemical compound produced by one microorganism which inhibits or kills other microorganisms.

Antiseptic: Substance that prevents the growth of bacteria.

Apiculus: A minute projection at the basal end of the spore by which the spore is attached to the sterigma.

Apothecium: An open ascospore.

Ascocarp: A fruiting body of Ascomycetes containing asci.

Ascomycetes: Group of fungi producing their sexual spores; within the asci.

Ascus: A sack-like hyphae containing ascospores. which are usually formed as a result of karyogamy and meiosis in ascomycetes fungi.

Aseptic: Sterile condition: no unwanted organisms present.

Autoclave: A container or any form of pressure cooker in any dimension (small or big) the contents of which can be heated up to 121 °C. It must be able to withstand an overpressure of 1 bar; otherwise the temperature cannot rise sufficiently.

Bacteria: Microorganisms that may cause contamination in culture work. Grain spawn is very easily contaminated with bacteria.

Basidiocarp: A fruiting body that bears basidia.

Basidiospore: A spore borne on the outside of basidium; following karyogamy and meiosis.

Basidium: A club-like structure on which a definite number of basidiospores are borne (four).

Biological effieceincy (BE): One way to express the productivity of a substrate. Be = lbs of fresh weight of mushrooms/ lbs of dry weight of substrate at spawning time. The range of be for a commercial farm varies between 60-120%.

Blended compost: a mixture of wheat straw bedded horse manure and other materials such as hay, wheat straw, corncobs, cotton seed hulls, etc. In several formulations, i.e., 80% h.m., 20% hay and cobs, etc break: see crop, cropping cycle.

Blotch: A disease characterized by large spots or blots which are irregular in shape on leaves, shoots and stems.

Brewers grain: grain hulls, residue from breweries, having a nitrogen content of 4.0 to 4.3%. Buttons: marketable mushrooms, but not mature.

Button stage: The young mushrooms are still fully closed.

Canning: Sealing in airtight containers freshly cooked food.

Carbon dioxide (CO_2): a by-product of the microorganisms during fermentation in both phase i and ll. It is also a very important by-product of spawn run, casing and later in production.

Casing inoculum (CI): low nutrient materials like vermiculite, peat or spent mushroom substrate that is sterilized and then it is fully colonized with mushroom mycelium. This material is then added to the casing layer to speed up the colonization of the casing layer and shorten the time to harvesting. Fully colonized spawn run compost may also be used but must be free of all molds or potential pathogens. Adding this compost at casing (cac) is not as commonly used as ci.

Casing layer: the top-dressing is required to induce the fruiting of the mushroom mycelium. Peat moss with limestone is main casing material used world wide. Spent mushroom substrate and loam top-soil can also be used.

Casing: the casing operation is the fourth step in mushroom farming and is a topdressing placed directly on spawn- run compost from 14 to 21 days after the spawning operation.

Cellulose: An organic compound in wood, straw, etc. It is easier to degrade than lignin. Cellulose is probably best known as raw material to make paper. Cotton waste contains high amounts of cellulose; sawdust contains cellulose, hemi-cellulose and lignin. Conditioning Gradual lowering of the temperature within one or two days in order to get rid of the free ammonia in the compost.

Colonization: Growth of mycelium in the tissues of the substrate or host.

Colonize: the process of the thread-like strands of growth, called mycelium, that develop in the compost after having grain spawn applied. Competitor molds: any of a variety of molds, when present in the compost or casing, which compete for nutrients, inhibits or destroys mushroom mycelium.

Compost turner: a machine specially designed to manipulate the raw materials (compost) in phase i into a homogenous rick or pile.

Composted substrate (compost): a mixture of organic and inorganic substances, managed specifically to produce nutrients (food) selectively, favorable to the growing of the common cultivated mushroom.

Cropping, cropping cycle: the sixth step in mushroom fanning begins 16-20 days following casing when the first mushrooms are harvested. Mushrooms are generally harvested (picked) for 3 to 5 days, followed by several days when no mature mushrooms are present. The period between harvesting is used for watering the casing layer. This cycle is repeated in a rhythmic fashion for the duration of the crop and is also called a break or flush.

Culture medium: Microorganisms differ in their nutritional needs. A large number of different media have been developed; PDA-agar and Malt-agar can be used for most cultivated mushrooms.

Culture: See mother culture.

Cyanophilic: Spore wall and ornamentation readily absorbing blue stain.

Cystidiole: A sterile cell in the hymenium situated between the basidia, arising from the same hymenial level as the basidium but differing from it in shape.

Deadrophysia: Much branched, often dendriform cell found in the hymenium or in the cap cuiticle.

Decurrent: Lamella which runs down the stipe.

Dimidiate: Gills which reach half way to stipe or pileus with semicircular outline.

Disinfect: To kill pathogen within the tissue or part of plants.

Double: the term used to indicate a standard mushroom-growing house, usually but not necessarily having 8,000 square feet of surface growing area.

Dry-matter (expressed in percent): Pertains to the amount of compost sample remaining after having been dried in an oven at 100°C for a 24-hour period. Precise weighing and handling is required to assure accurate data.

Eradication: Control of plant disease by eliminating the pathogen after it is established or by eliminating the plants that carry the pathogens.

Etiology: The study of cause of a disease or disorder.

Fermentation: The process of composting. Easily accessible nutrients will be degraded by microorganisms that make the substrate more selective. Unwanted fermentation may occur if the compost is still very 'active' or if thick layers or large bags are used. In that case the temperature rise inside the substrate will become too high for the desired mycelium.

Fertilization: The sexual union of two protoplasts resulting in doubling of chromosome numbers.

Filamentous: Growth composed of long, irregular placed or interwoven threads.

Filling: the moving of compost from phase i into containers or a structure to undergo phase ii composting

First break: the time when the first mushrooms of each crop are harvested, usually 16-20 days following casing.

Flush: The sudden development of many fruiting bodies at the same time. Usually there is a resting period between flushes or breaks.

Flushing: another important step required to promote the development of mushroom initials. Fresh air is introduced to reduce the level of carbon dioxide produced by mushroom mycelium. Temperature and relative humidity is also adjusted to move or hold back pin development.

Formol: A 30% solution of formaldehyde used to sterilize areas. The gasses kill living microorganisms and spores.

Free water: The actual water available to the microorganisms in the substrate. Water content is the absolute measure. Free water is related to the water film around each particle in the substrate and the concentration of salts in the water.

Friable: the physical structure and condition of loam top-soil, i.e., crumbly.

Fruit body: A part of the thallus differentiated in shape, consisting of hyphae and reproductive spore producing cells.

Fruiting: The mycelium will form mushrooms in its reproductive stage. This is called fruiting as the mushrooms are actually the fruiting bodies of the mycelium.

Fumigation: Application of fumigant for disinfection.

Fungiside: A substance that kills fungal spore or mycelium.

Germ tube: The early growth of mycelium produced by a germinated fungus spore.

Germination: The spreading of hyphae from spores.

Gills: the radially arranged, vertical plates below the cap of a mushroom on which spores are formed. Hyphae, hyphae: Individual cells of mycelium.

Glebs: The fruiting body tissue of the Gasteromycetes producing basidia.

Gypsum, agricultural: a naturally occurring mineral consisting of calcium sulfate. It is used in phase i composting to prevent greasiness.

Halobasidium: A single-celled basidium, although typically club-shaped, holobasidia may resemble turning forks in some taxa, while in others the basidium may become divided by adventitious septa.

Haploid: A cell ore organism whose nuclei have a single complete set of chromosome.

Homogenous mix: The thorough and complete mixing of all materials used.

Horse manure: one of the basic bulk materials used mushroom growers for substrate preparation. Wheat-straw bedding, containing very little hay, wood shavings, sawdust, or other foreign bedding materials is used. Horse manure bedded on nonfibrous material is not usable as a bulk ingredient. Hyphae: single strands of mushroom mycelium.

Hybrid: The offspring of two individual differing in one or more heritable characters.

Hymeniderm: The structure of the hymenidermal cuticle of the cap is similar to that of the hymenium.

Hypha: A single tubular filament of mycelium.

Incubation: The period after inoculation (preferably at an optimum mycelial growth temperature) during which the mycelium slowly grows through the substrate

Inhibition: Prevention of growth or multiplication of microorganisms.

Inner veil: The hyphal membrane which covers the gills of a young mushroom.

Inoculation: Transferring an organism into a specific substrate.

Isolate: a product of a tissue, multi-spore or single spore culture maintained on a nutrient medium.

Lignin: An organic substance that is difficult to degrade, which, together with cellulose, forms the basis of wood, straw, etc.

Line: a mushroom isolate maintained over time.

Mesophilic: The first microorganisms to become active in phase i compost and as they grow and multiply causing temperatures to increase rapidly. These initial mesophilic organisms are incapable of growth at temperatures exceeding 45°c (112°f).

Microorganisms: Microscopic organisms which are abundantly present in the air and stick to every surface.

Microorganisms: the smallest form of living organisms (or animal life) found in most raw materials and can only be seen by use of a microscope.

Moisture: in mushroom growing, refers to a certain amount of water diffused in the compost, casing layer, etc. Moisture is one of the most important, measurable elements to be monitored during the entire mushroom growing process.

Mother culture: A pure strain of an edible fungus growing on a medium.

Mother spawn: Spawn that is not used for inoculating substrate, but for inoculating another batch of spawn.

Mould: Any profuse or woolly fungal growth on damp or decaying matter or on surfaces of plant tissue.

Muriform: Spores having septa in more than one place.

Mushroom cap: The very top portion of the mushroom, sitting (attached) on top of the stipe (stem). The cap is the only part of the mushroom that may have a different color as related to the 'line' being grown.

Mushroom mycelium: the white or grayish white, thread-like growth that develops after spawning. It is the conduit through which the mushroom gets its nourishment from the compost. Nitrogen: an important and measurable element of substrate. Results on a dry weight basis are used by growers to monitor supplementation during phase i substrate preparation and at spawning for efficiency of phase ii composting.

Mushroom: fruit body of the mycelium at the surface of the casing layer. Sporophore is another term for mushroom fruiting body.

Mutant: An individual possessing new, heritable characters as a result of a mutant.

Mycelium: The network of hyphae that form the vegetative body of the fungus. Mushrooms are the fruiting bodies of the mycelium.

Mycorrhiza: A symbiotic relationship between fungi and plant roots.

Nutrients: those ingredients added at the beginning of phase i composting that are directed at feeding the microbial population. Nutrients added at spawning are directed toward actual consumption by the mushroom itself.

Ostiole: A neck-like structure in an ascocarp; lined with periphyses and terminating in a pore, also the opening of a pycinidium.

Parasite: Organism that lives at the expense of others, usually causing diseases in its hosts. Ultimately it may cause the death of its host.

Pasteurisation: Heat treatment applied to a substrate to destroy unwanted organisms but keeping favourable ones alive. The temperature range is 60-80°c. The treatment is very different from sterilisation, which aims at destroying all organisms in the substrate.

Peak Heating: Pasteurisation of the compost in the growing rooms Pinhead: A term to describe a very young mushroom, when the cap is pin-sized.

Pests, mushroom: all unwanted organisms which interfere with mushroom production, that by infestations, compete for nutrients needed by the mushroom, or directly attack the mycelia strands or the mushroom itself. Example pests are nematodes, mushroom flies, virus diseases, mummy, fungal pathogens, competitor molds.

Petri dish: A round glass or plastic dish, with a cover, used for observing the growth of microscopic organisms. The dishes are partly filled with sterile growth medium (or sterilised after they have been filled). Petri dishes are commonly used to grow mycelium that will inoculate the mother spawn.

Phase I composting: the first step of six steps involved in mushroom farming. This first step is to create compost with sufficient nutrients to grow mushrooms and at the same time, provide little or no nutrition for other fungi or competitor organisms. This process involves much higher temperatures than those required in phase ii composting.

Phase II composting: the second step in mushroom farming, after phase i composting is complete and the compost has been filled into containers or into a structure. Phase ii composting has two main purposes: formal conditioning of compost so it becomes mushroom specific (absence of ammonia and readily available carbohydrates) and pasteurization. This step is performed under controlled environmental conditions and at much lower temperatures than phase i composting. See pasteurization.

Pileus: Upper portion of cap or certain types of ascocarps and basidiocarps.

Pinning: the fifth step in mushroom farming and is initiated when rhizomorphs form in the casing and then emerge at the surface of the casing. There are two important steps involved in causing the mycelium to go from the vegetative stage to the fruiting stage. The two steps are to lower the air temperature approximately 10°f and introduction of outside fresh air to purge the CO_2 from the surface of the casing. At the time of pinning, a final watering is sometimes added to protect against drying out of the casing layer due to the introduction of fresh air.

Plasmogamy: The fussion of two protoplasts.

Pleurocystdium: A cystidium occurring on the face of gills.

Poultry manure: phase i compost supplement used in two forms, i.e., mechanically dried & pulverized or raw, uncured. In either form it is used as a source of nitrogen at the beginning of phase i composting. Nitrogen content varies but should be approximately 4%.

Pre-wet turner: a machine specially designed to manipulate the raw materials (compost) in phase i into a homogenous material, usually in a windrow not rick primordia: see mushroom initials.

Primordium: The initial fruiting body.

Production cycle: the total time of mushroom growing beginning with build through to steam-off and clean-out

Propagule: It is that part of the fungus by which it may be dispersed or reproduced.

Pure culture: An isolated culture of a microorganism without any other microorganisms. Pure cultures are essential to the spawn production process.

Radial: Spreading out radially from the control to the margin.

Relative humidity: The percentage of moisture in the air compared to the maximum amount of moisture that the air can contain at a given temperature and pressure.

Rhizomorphs: the visibly thick strands or cord of compacted mycelia which is the stage between the mycelium and mushrooms. Mushrooms actually form from rhizomorphs.

Rick (windrow, pile): phase i composting term for the windrow of homogenous mix of raw materials used to produce mushroom compost. The rick has straight and tight sides and looks like a loaf of bread. Rhythm: refers to the cycles that develop after the first mushrooms are harvested.

Rose face: a specially handcrafted and drilled device used for applying water to the casing layer from the day of casing and throughout the cropping period. Rose faces come in various sizes, hole sizes, that are selected by each mushroom farmer according to his needs or experience with his particular casing material.

Rot: The softening, discolouration and often disintegration of a Succulent plant tissue as a result of fungal or bacterial infection.

Sanitation: The removal and burning of infected plant parts, decontamination of tools, equipments, hands, etc.

Saprophytes: fungi that break down complex organic structures of plants and animals in order to feed on them

Saprophytic mushrooms: e.g. *Agaricus* spp and *Volvariella* spp.

Sclerotium: A compact mass of hyphae with or without host tissue, usually with a darkness rind and cabable of surviving under unfavorable environmental conditions.

Scratching: not a recommended practice, however, it is performed by loosening the casing layer by home-made devices, usually a narrow board with nails driven through it. Scratching is normally done to a depth just short of where the growth of mycelium is observed. The reasons mushroom farmers perform this extra step varies, however, sealing is the normal cause. Sealed: an undesirable condition of the casing layer produced by too much water applied at one time to a particular area, or with too much force. This condition exists when particles of the casing soil re-align themselves in a pattern that drastically reduces the surface pore space. Very few, if any, mushrooms are found on a sealed casing layer.

Slant culture: A small culture consists of an organism as distinguished from its reproductive organs or reproductive phase.

Slant: A test tube with growth medium, which has been sterilised and slanted in order to increase the surface area.

SMS: Spent Mushroom substrate, the substrate remaining after the mushrooms have been harvested.

Somatic: The body phase-in plants; the vegetative phase-structure, or function as distinguished from the reproductive phase.

Spawn run: The period of vegetative growth of the mycelium throughout the substrate after spawning.

Spawn: Mycelium growing on a substrate used as planting material in mushroom cultivation.

Species: Fundamental unit of biological taxonomy. Generally speaking, two individuals belong to the same species if they can produce fertile offspring.

Spores: mushroom spores are produced on the gills of the cap of the mushroom. They are microscopic spheres roughly comparable to seeds of higher plant life. Spores are ~ used to spawn mushroom compost. Sporophore: see mushroom.

Square feet: the unit of measurement used in most calculations or determinations in mushroom culture. It is a reference to surface area (length times the width) which is used from the inception of building a new mushroom farm, to how much pesticide to apply, to how a crop is yielding, i.e., pounds of mushrooms picked per square foot of surface area.

Stem: stalk of a mushroom

Sterigma: A small slender protuberance which supports a sporangium, a conidium or a basidiospore.

Sterile: Conditions: see aseptic.

Sterilisation: Destroying (completely) all microorganisms present, by heat. Spawn substrate always has to be sterilised prior to inoculation.

Sterilize: in mushroom farming, all equipment, such as utensils, knives, etc., that will come in contact with the mushroom or substrate after phase ii, must be sterilized to clean and free it from harmful pathogens. Note: compost is not sterilized in phase ii.

Stipe: Stalk of a mushroom.

Strain: A group of individuals within a species. Equivalent to "race" or "variety" in plants. Subculture: A culture derived from another culture.

Stroma: A rigid structure composed of hyphae where in fruit bodies or cavities containing asci or conidiospores are embedded.

Substrate: the material in which the mycelium grows.

Supplements: in phase i composting: standard supplements include dried or raw poultry manure, brewer's grain, cottonseed meal, agriculture gypsum, cottonseed hulls, crushed corn cobs, cocoa shells and possibly others, depending on the recipe. See nutrients. After phase ii, supplements include cracked soybean, corn by-products that are treated to delay the release of the nutrients so that they are available to the slower growing mushroom mycelium and not available to the faster growing weed molds.

Syndrome: A group of signs and symptoms that occur together and characterize the disease.

Synthetic compost: compost made with bulk ingredients other than horse manure. Various recipes are used where straw, hay, are supplemented with supplements like crushed corn cobs, cottonseed hulls, poultry manure and sometimes are inorganic forms of nitrogen (e.g. Urea, ammonium nitrate) are used.

Taxonomy: Science of classification of living things Test tube: A tube of thin, transparent glass closed at one end used in chemical and biological experiments.

Thallus: A relatively simple plant body devoid of stems, roots and leaves; the somatic phase in fungi.

Thermocouple: a special temperature sensing device, consisting of two dissimilar metals, manufactured in a stainless steel sheath. These devices require special electronic equipment to accurately read whatever medium they are placed in. Thermocouples are normally used in phase i, phase ii or phase iii as well as any place where remote reading of temperatures are necessary or advantageous throughout the cropping cycle.

Thermophiles: these are heat loving organisms that survive and multiply in temperatures up to 150°F. See microorganisms. Turning: occurs during phase i composting and after building a rick or in a bunker. It means to tear apart the rick, mixing the colder outside with the hotter inside compost, while adding supplements or water or neither, depending on each composter's regimen.

Tissue culture: A culture made from the tissue of a young and healthy mushroom.

Vegetative: the growth of mycelium from spawning through to pinning. In order to cause fruiting (form mushroom initials), all vegetative growth must cease. See pinning.

Veil: Layer of tissue that completely surrounds the baby mushroom.

Ventilate: introduction of outside fresh air during phase ii composting, the fruiting or pinning process and throughout the cropping cycle.

Viable: Cabable of living, growing and developing alive.

Volva: A cup at the base of the stipe of certain much rooms; a remnant of the universal veil.

Yield: the pounds of mushrooms harvested per square foot. Mushroom growers manage and make economic and cultural decisions based on variations yield.

Zygote: A diploid cell resulting from the union of two haploid cells.

489